THERMAL ANALYSIS

VAN NOSTRAND REINHOLD
·SERIES IN ANALYTICAL CHEMISTRY

Edited by

DR. R. A. CHALMERS

Department of Chemistry
University of Aberdeen

———

This series is designed as a coverage of reliable analytical information of value
to chemists in research, industry and teaching. Each volume is carefully
selected and planned as a modern treatment of a topic of importance to analytical
chemists today. New volumes will be added from time to time.

Additional titles will be listed and announced as published

THERMAL ANALYSIS

by

RNDr. PhMr. ANTONÍN BLAŽEK, C.Sc.

Senior Scientist, Laboratory of Thermoanalysis,
Central Laboratories, Institute of
Chemical Technology, Prague

Translation Editor:

JULIAN F. TYSON, B.Sc.
Imperial College of Science and Technology,
University of London

VAN NOSTRAND REINHOLD COMPANY
LONDON
NEW YORK CINCINNATI TORONTO MELBOURNE

VAN NOSTRAND REINHOLD COMPANY LTD
25—28 Buckingham Gate, London S.W.1

INTERNATIONAL OFFICES
New York Cincinnati Toronto Melbourne

COPYRIGHT NOTICE

© Antonín Blažek 1972

Translated by Želimír Procházka
First published in 1973
by Van Nostrand Reinhold Company Ltd,
25—28 Buckingham Gate, London S.W.1
in co-edition with SNTL-Publishers of
Technical Literature, Prague

© Van Nostrand Reinhold Company Ltd, 1973

Library of Congress Catalog No. 70-188528

ISBN 0 442   00812 0

Printed in Czechoslovakia by SNTL., Prague

CONTENTS

PREFACE

Thermal analysis, including, mainly, thermogravimetry (TG) and differential thermal analysis (DTA), is widely used in chemical laboratories. In the past twenty years the equipment for these methods has been extensively developed, and is now readily available commercially. In view of the possibilities offered by such methods in the investigation of both physical phenomena (e.g. changes in crystallographic properties, melting, sublimation, adsorption) and chemical phenomena (e.g. dehydration, decomposition, oxidation, reduction), methods of thermal analysis have found application in almost all the natural sciences, including chemistry, geology, mineralogy, and metallurgy. A detailed study of these methods has shown that they have their limitations, and that the results obtained by various workers on analysing the same material do not always agree. The study of experimental conditions and the construction of apparatus has helped to elucidate the differences in the results obtained and improved the applicability of both methods.

Only two monographs dealing with the problems of thermal analysis are to be found in the Czech specialized literature. However, review articles and original papers from this field are quite numerous, indicating the great interest in these methods in Czechoslovakia. There are several monographs in the world literature on the historical development, theoretical principles, apparatus, and a number of applications of the methods of thermal analysis.

This book is intended to cover the two fundamental methods of thermal analysis (TG, DTA), the possibility of using them in qualitative and quantitative analysis, the study of reaction kinetics, and the factors which limit the reliability and the applicability of the results obtained, especially in quantitative evaluation. The description of some apparatus (mainly commercial) is intended to illustrate the development of and current achievement in instrumentation. In view of its size the book cannot cover the given theme in full detail, and the reader is advised to consult the following books for a more thorough insight into some particular branches of the method (the references are at the end of Chap. 1): the book by C. Duval [18] for apparatus, TG and automatic gravimetric analysis; the

3

books by W. J. Smothers and Y. Chiang [69], R. C. Mackenzie [44, 44a], and D. Schulze [93] for DTA apparatus and its applications; and the monographs by P. D. Garn [25] and W. W. Wendlandt [86] for more detailed theory and experimental conditions of both methods. A theoretical treatment of the DTA curve can be found in the book by G. A. Piloyan [59]. It is also necessary to stress that the number of new publications on this topic (especially on the applications of these methods), which appear every year is rather large, and that the size of this book does not allow them to be discussed extensively. In this respect, the reports of various international conferences on thermal analysis [16, 47, 61, 95, 97], as well as the regularly published reviews on progress in this field [2, 50, 51–53, 62, 98, 99, 100–103] are recommended. The number of papers on thermal analysis increases every year. All are systematically reviewed in "Thermal Analysis Review" [62] which started in 1964. Every two years reviews on these papers appear in "Analytical Chemistry" [50–53, 96]. In 1969 a new journal, "Journal of Thermal Analysis", appeared, devoted to papers on thermal analysis [32], and two more books were published by C. J. Keattch [34] and M. Harmelin [28]. The Elsevier Publishing Co. has issued a new journal, "Thermochimica Acta". Some firms producing thermobalances and equipment for DTA also issue information bulletins on the work performed with their apparatus. These include, for example, the firms Linseis in the German Federal Republic, Mettler in Switzerland, Shimadzu in Japan, Perkin-Elmer and Du Pont in the U.S.A., and Stanton Redcroft in the U.K. This abundant publishing activity shows that the importance of the methods described is steadily increasing.

Tables summarizing TG and DTA data are an integral part of this book. Complete tabular presentation of the published results would be beyond the scope of this book and, therefore, the review has been limited to those from certain interesting fields. For more detailed information on thermogravimetric data the reader should consult the book by C. Duval [18], and the punched-card index of data on DTA of mineral, inorganic, and certain other substances [45], compiled by R. C. Mackenzie. The Sadtler Company of the USA also sells a similar index of data covering a much larger area of substances, including commercially produced pure organic and inorganic compounds, steroids, and drugs [92]. In addition to this, a number of publications contain compilations of data on thermal analysis of various materials from single areas of investigation [23, 36, 64, 94].

The author gratefully acknowledges permission to reprint figures and tables, so generously and readily granted by the copyright holders.

GLOSSARY OF SYMBOLS

a	thermal diffusivity ($\lambda \cdot c^{-1} \cdot \varrho^{-1}$)
A	constant (of the apparatus)
$B, \Delta B$	area of effect on the DTA curve
c, c_p	heat capacity, specific heat at constant pressure
d, ∂	differential
E	activation energy
G	Gibbs free energy (ΔG, change of free energy)
g	constant (effect of geometrical arrangement on heat transfer in DTA)
h	height of a sample of cylindrical shape
H	enthalpy (heat content)
ΔH_0	heat of reaction under standard state conditions
I	shape index of DTA curve
K_p	equilibrium constant of reaction
k	Arrhenius rate constant $= Z \cdot \exp(-ER^{-1}T^{-1})$
M, M_a	mass of the reacting substance (active mass)
n	order of reaction
P	gas pressure
$Q, q, \Delta Q$	heat of reaction, heat of transformation
R	gas constant
r	radius of the sample
S	entropy
T	temperature
ΔT	temperature difference
t	time
V	volume
Z	frequency factor
α	degree of transformation of the reacting substance
$d\alpha/dt$	rate of transformation (reaction)
ω	weight of the sample
$\Delta \omega$	change of weight
ϱ	density
λ	coefficient of thermal conductivity

∇	Laplace operator
Φ	rate of heating, dT/dt

INDICES:

sm	sample
st	reference material (standard)
m	maximum of curve
o	point of origin
i	given medium
1, 2, 3, ...	characteristic points on curve
s	steady state
c	block

ABBREVIATIONS:

TG	thermogravimetry
DTA	differential thermal analysis
DTG	derivative thermogravimetry
DDC	dynamical differential calorimetry
DSC	differential scanning calorimetry, enthalpic thermal analysis
FTA	fractional thermal analysis
e.m.f.	electromotive force

Chapter 1.

INTRODUCTION

Methods of thermal analysis are a related group of techniques whereby the dependence of the parameters of any physical property of a substance on temperature is measured [82]. The majority of them follow changes in some property of the system (mass, energy, dimensions, conductivity, etc.) as a dynamic function of temperature. The basic parameter important for the methods of thermal analysis is the change in heat content (ΔH). Every substance can be characterized by its free energy (G), given by the expression $G = H - TS$, where H is enthalpy, T is absolute temperature and S is entropy.

At a given temperature, every system has the tendency to attain a state in which the free energy is at a minimum, for example, the transition from one crystalline form of a substance to another which, at a given temperature, has a lower free energy and is therefore more stable. The formation of a more stable crystalline structure, or of another state with a lower free energy, may take place on gradually heating the sample, or via intermediary steps. For example, melting, boiling, sublimation, change of crystalline structure, chemical reaction, etc. may represent such a transformation. The transformation is characterized by the temperature at which it occurs and by a change in heat content, manifested by an increase or decrease in the temperature, depending on whether the reaction is exo-or endothermic. This is the basis of differential thermal analysis (DTA).

The change in heat content can also be accompanied by a change in weight, e.g. during a chemical degradation, dehydration, sublimation or oxidation. Observation of such a change is the basis of the thermo-gravimetric method (TG).

When a substance is heated or cooled, reversible or irreversible changes in its dimensions take place, depending on the initial dimensions and on the temperature. Thermal expansion is a result of increasing vibration of atoms around equilibrium positions in the crystal. The observation of the change in dimensions during heating is the basis of dilatometry, and has great importance in metallurgy, physics and glass and ceramics technology. The coefficient of expansion of crystalline substances and the resulting change in dimensions may also be observed by electron or X-ray diffraction techniques, which determine the exact lattice parameters. In addition, the changes in characteristic X-ray reflections may serve to

Table 1.1

Methods of Thermal Analysis

Technique	Quantity Determined	Apparatus
1. Thermogravimetry (TG)	Weight change	Thermobalance
2. Derivative thermogravimetry (DTG)	Rate of weight change	Thermobalance or derivative thermobalance
3. Differential thermal analysis (DTA)	Temperature difference between the sample and the reference substance	DTA apparatus
4. Derivative DTA	Derivative of the temperature difference	DTA apparatus
5. Differential scanning calorimetry (DSC)	Amount of heat transmitted to the sample	Differential calorimeter
6. Measurement of specific heat	Specific heat	Calorimeter
7. Evolved gas analysis	Amount of gas liberated	Gas analyser
8. Pyrolysis	Product of pyrolysis	Gas chromatograph, mass spectrometer, IR spectrometer, etc.
9. Thermal luminescence analysis	Light emission	Photomultiplier thermoluminescence apparatus
10. Dilatometry	Change in volume	Dilatometer
11. Electric conductivity analysis	Change in electric resistance	Resistance bridge
12. High-temperature X-ray diffraction	Change in lattice dimensions	X-ray diffractometer
13. Thermometry	Temperature change as a function of time or volume of the titration reagent	Thermometric titrator
14. Enthalpimetry	Enthalpy change as a function of amount of reagent added	Thermometric titrator
15. Classical calorimetry	Temperature as a function of time. Heat content as a function of temperature	Calorimeter

identify the reactions under investigation. The analysis of gaseous products of chemical reactions, [evolved gas analysis (EGA) and evolved gas detection (EGD)], and the measurement of other physico-chemical properties, e.g. electrical conductivity, thermal conductivity, optical properties, dielectric constants, thermoelectric potentials, and magnetic properties, are also the basis of other techniques which may be included in methods of thermal analysis, on the basis of the definition above. In Table 1.1 the commonest methods of thermal analysis are listed, together with an indication of the physico-chemical properties followed as the temperature is varied. The fact that chemical reactions, or changes of phase, are accompanied by a simultaneous change of several physico-chemical parameters (as for example the change in weight, ΔH, volume, liberation of gas, etc.) led to a combination of several thermoanalytical methods, resulting in a larger number of complementary results. The type of thermoanalytical method used depends on the information required from a particular experiment. If the aim of the experiment is the determination of one change, only the methods which are suitable for the given type of material and the change under investigation will be used. The observation of only one temperature-dependent property usually does not suffice for a definition of a chemical reaction, and a combination of several methods may be required. It should be pointed out that the agreement of the results obtained by various methods, and often by the same methods performed by different people, is sometimes very poor; this is due primarily to the great difficulty in obtaining identical experimental conditions. Suitable experimental conditions and the factors limiting the scope of these methods will be discussed in greater detail in appropriate chapters. With a view to improvement in the correlation of the results obtained by single methods, a combination of methods may be recommended, such as, for example, simultaneous TG and TDA, although each such combination represents a certain compromise in the choice of optimum conditions for each method.

A detailed description of each thermoanalytical method mentioned in the preceding section and Table 1.1 exceeds the scope of this book, which is limited to the two most important and commonly used methods, i.e. TG and DTA. These two methods have been widely used in the past 20 years, although they originated much earlier. This is due mainly to the fact that apparatus has attained a high technical standard during this period and is now readily obtained.

Methods of thermal analysis associated with a change in weight can be placed in two classes: static and dynamic. These have been defined as follows [82].

Static methods

Isobaric weight-change determination. A technique of obtaining a record of the equilibrium weight of a substance as a function of temperature (T) at a constant partial pressure of the volatile product(s). The record is the isobaric weight change curve; it is normal to plot weight on the ordinate with weight decreasing downwards and T on the abscissa increasing from left to right.

Isothermal weight-change determination. A technique of obtaining a record of the dependence of the weight of a substance on time (t) at constant temperature. The record is the isothermal weight change curve; it is normal to plot weight on the ordinate, with weight decreasing downwards and t on the abscissa increasing from left to right.

Dynamic methods

Thermogravimetry (TG). A technique whereby the weight of a substance in an environment heated or cooled at a controlled rate is recorded as a function of time or temperature. The record is the thermogravimetric or TG curve; the weight should be plotted on the ordinate with weight decreasing downwards and t or T on the abscissa increasing from left to right.

Derivative thermogravimetry (DTG). A technique yielding the first derivative of the TG curve with respect to either t or T. The curve is the derivative thermogravimetric or DTG curve; the derivative should be plotted on the ordinate, with weight losses downwards, and t or T on the abscissa increasing from left to right.

Inverted thermogravimetry. More recently, this technique has been described, in which the substance under investigation is heated and the volatile products are trapped on an adsorbent (e.g. charcoal) which is weighed continuously [77a].

Differential thermal analysis (DTA) is a technique of recording the difference in temperature between a substance and a reference material against either time or temperature as the two specimens are subjected to identical temperature regimes in an environment heated or cooled at a controlled rate [82]. The results may be either qualitative or semiquantitative. The record is the differential thermal or DTA curve; the temperature difference (ΔT) should be plotted on the ordinate with endothermic reactions downwards and t or T on the abscissa increasing from left to right. Both basic methods of thermal analysis have been used for a long time. The idea of following the course of chemical reactions as a function of time or temperature by discontinuous weighing is so simple and widely known that it is very difficult to indicate the exact date of origin of the thermogravimetric method. A detailed discussion of the historical development of the method, mainly from the point of view of experimental technique, is given by Duval [18, 19] and by Keattch [34]. The first use of this technique is mentioned in papers from 1893 and 1914, although the equipment used at that time was not called a thermobalance. If we define the thermobalance as an apparatus permitting the continuous graphical representation of changes in weight of the sample as a function of temperature or time, while it is being heated or cooled, we can look for the date of origin of this method in the period from 1905 to 1923. In a series of papers from this period [30, 54, 83] an apparatus was used for which the Japanese author Honda in 1915 coined the name "thermobalance". Thermogravimetric equipment has evolved from the original primitive device which could usually only be used under atmospheric conditions, to the present advanced and highly technical apparatus. The merit for this successful evolution belongs mainly to the French school, represented by Guichard, Chevenard, and Duval. The thermobalance constructed by Chevenard was the first commercially produced apparatus of this type in the world. In 1953 Duval published a book, extensively revised in the second edition [18], in which he described a method of using thermogravimetry in analytical chemistry (so-called automatic gravimetric analysis) and gave thermogravimetric curves of a large number of inorganic analytical precipitates. The development of thermogravimetry in Czechoslovakia is due to S. Škramovský and his co-workers, who, as early as 1932, described the construction of a thermobalance fitted with a photographic recorder. He gave this method the name "stathmography" [78]. [This name is derived from the Greek noun stathmos (balance, weight) and thus has a broader meaning than the term "thermogravimetry", which is limited to thermolytic processes.] Škramovský and his co-workers worked mainly on kinetic studies and the investigation of the stability of various hydrates. As early as 1934–1937 [79, 80], these authors indicated the possibility of erroneous evaluation of TG curves e.g. from apparent shifts in inflexion points, arising from kinetic effects. Their work has shown that during the dehydration of substances such as $CuSO_4 . 5 H_2O$, and the decomposition of carbonates of bivalent metals at elevated temperatures, the rate of decomposition follows $1/3$ order kinetics. The development of the methods of thermal analysis in Czechoslovakia was also furthered by the work of Bárta and his co-workers. The organization of several conferences on thermal analysis in Prague is due to the efforts o this group, which worked at the Department of Silicates of the Institute of Chemical Technology. These conferences contributed substantially to the spread of this method. In Czech journals a series of papers can be found in the period after the second World War which describe not only new thermoanalytical apparatus, but also many applications of TG and DTA in the fields of chemistry, metallurgy, mineralogy, and several others. [1, 2, 4–13, 17, 22, 24, 29, 31, 60, 65, 70–77, 81, 84, 85, 89].

The beginnings of differential thermal analysis may also be sought in approximately the same period as those of the thermogravimetric method. In 1886 Le Chatelier used this method for the first time [37–41] for the study of calcite and later for the study of clay materials. At its beginnings, of course, this method consisted of a direct determination of the rate of change in temperature of the sample during regular heating. The reactions followed in this way gave a series of plateaux on temperature versus time plots, and their determination was rather inaccurate. The method was considerably improved in 1891 by Roberts-Austen [63, 64] by the introduction of a differential thermocouple which measured the difference in voltage between thermocouples placed in the sample and in an inert standard. This system was then gradually improved [15, 43, 66] and introduced in other fields of metallurgy. As a result of the papers by Kurnakov, Rode, and Geld, this method, and thermal analysis in general, spread appreciably in the Soviet Union, where the first "thermographic" conference [90] was held in 1953. As the DTA method developed, it became more extensively used in all fields of natural science. From the original devices with manual or photographic recording of curves, development led to fully automatic apparatus with direct recording. Advanced apparatus for qualitative and quantitative analyses can be found on the world market, and also special apparatus for calorimetric measurement. A detailed review of the development of this experimental method, and of the results attained in various fields can be found in the monographs by Smothers and Chiang [69], Mackenzie [44, 44a], and others.

Both DTA and TG are widely used in situations where physico-chemical properties accompanied by changes in the heat content and the weight of the investigated material are studied. Results can be evaluated qualitatively and quantitatively both in analysis, and for obtaining thermodynamic data, e.g., in studies of reaction kinetics. The application of thermoanalytical methods is widespread and several broad reviews have appeared [18, 43, 57, 62, 90]. Although both methods are utilized in almost all branches of chemistry, they are mainly employed in analytical, inorganic, organic and physical chemistry. Table 1.2 lists some typical phenomena which can be investigated by these methods. However, it should not be forgotten that both methods have certain limitations which complicate the interpretation of results. Often information obtained by these methods is of a purely empirical nature. For example, in TG, information on the temperature at which the curve departs from the base line, and in DTA. the temperature at which the maximum heat effect occurs, were considered as very precise and important for the definition of the temperature of the reactions investigated. In actual fact, these values depend on a series of factors, the analysis of which will be given in subsequent chapters. The development of both methods has led to the construction of various types of apparatus which very often differ in the definition of experimental conditions. This means that the comparability of published results is sometimes poor. This is evident from the poor correlation between the results obtained by the DTA and TG methods separately, and also from the comparison, carried out by Mackenzie and Farquharson [46], of the application of DTA to samples of standard minerals made by various authors; the differences in the temperatures given for the peaks of the heat effects were often greater than 100 °C.

A particular worker may produce results which agree well amongst themselves, but are quite different from those of others. Assuming that under identical thermodynamic conditions, samples of the same material will behave identically, the reason for this discrepancy in the behaviour of the samples should be sought in differences in the experimental conditions, i.e. in differences in technique. Differences in experimental conditions may lead to differences in the indication of temperature of the reaction; for example, the composition of the atmosphere in the furnace may substantially change the course of the reaction. The first papers to analyse these effects on the possible distortion of thermal analysis curves appeared in the years 1954–1958 [15, 20, 35, 48, 67]. Later on, considerable attention was devoted to this question [26, 55], and various methods of correction such as empirical correction equations and correction curves were proposed [58, 67, 87]. From these papers it is evident that for the objective evaluation of the results of dynamic thermoanalytical methods, simultaneous introduction of corresponding experimental data is necessary. This is true both of TG and DTA.

At the first international conference on thermal analysis, held in Aberdeen in 1965 [61] and organized by the International Committee for Thermal Analysis (ICTA), a standardization committee was nominated. This had the duty of checking the possibility of standardizing the methods used and proposed the principles for the publication of results [88]. This committee concentrated on apparatus for thermal analysis, and the actual identification of materials, experimental conditions, reproducibility of recordings, etc. Keattch [33], a member of this committee, proposed 31 compounds for the standardization of temperature in thermogravimetry. The methods of temperature measurement in TG and DTA, as well as the utilization of temperature and heat standards will be discussed more thoroughly in the

Table 1.2

Some Applications of TG and TDA

Process investigated	Substances	Method
1. Dehydration—determination of free and bound water	Organic and inorganic compounds (precipitates, minerals, combustibles, coordination compounds, etc.	TG DTA
2. Thermal decomposition	Organic and inorganic substances (precipitates, catalysts, minerals, polymers, etc.)	TG DTA
3. Roasting and calcination	Minerals	TG DTA
4. Distillation and evaporation	Inorganic and organic substances	TG DTA
5. Thermal oxidation—corrosion	Inorganic and organic substances (metals etc.)	TG DTA
6. Solid phase reaction	Inorganic and organic substances	TG DTA
7. Gas-solid reactions (oxidation, reduction, corrosion)	Inorganic and organic substances	TG DTA
8. Study of new chemicals	Inorganic and organic substances	TG DTA
9. Catalysis	Inorganic and organic substances	TG DTA
10. Study of reaction kinetics and reaction mechanism	Inorganic and organic substances	TG DTA
11. Study of heats of reaction	Inorganic and organic substances	Calorimetry DTA
12. Thermal stability and purity	Analytical reagents	TG DTA
13. Chemical composition determination	Complex compounds	TG
14. Study of drying and annealing (combustion)	Analytical precipitates, filtration technique	
15. Development of gravimetric analytical methods	New analytical precipitates	TG
16. Automatic thermogravimetry	Simple compounds and mixtures	TG
17. Phase changes	Organic and inorganic compounds (glasses, liquid crystals, ceramics, minerals etc.)	DTA

appropriate chapter. An equally important problem in publishing thermal analysis results is the use of a unified and agreed nomenclature. Certain efforts in this direction may be traced to 1957 [3]. The nomenclature committee of ICTA, also nominated in 1965 at the Aberdeen conference, proposed in 1969 the abbreviations TG and DTA and recommended definitions for the two methods [82]. It is evident that considerable efforts are being made to standardize the reporting procedure of both methods and to improve the correlation of results. This is also evident from the fact that the standardization committee prepared an exhaustive report for the second international conference (Worcester, 1968), directed not only towards nomenclature problems and the compilation of a review of all the possibilities of thermal analysis (including apparatus and procedure), but also to the problems of measurement, working standards, and the correct method of evaluating temperature points and defining them. The chairman of the standardization committee is H. G. McAdie of Toronto, and 25 research laboratories from all over the world contribute to the work of the committee. The chairman of the nomenclature committee is R. C. Mackenzie of Aberdeen. The results of the work of both these committees will contribute to a further utilization of both methods and to an improvement in the usefulness of the results.

REFERENCES

1. BÁRTA R. *Silikáty* **6**, 125 (1962).
2. BÁRTA R., ŠATAVA V. *Stavivo* **15**, No. 1 (1953).
3. BÁRTA R. *Silikáty* **1**, 191 (1957).
4. BERÁNEK M. *Silikáty* **10**, 93 (1966).
5. BERÁNEK M. *Report of the Research Institute of Low-Tension Technique, Tesla*, Prague No. 13045/3, Prague 1963.
6. BERG L. G. *Vvedenie v termografiyu.* Izd. Mosk. Univ., Moscow 1958.
7. BERGSTEIN A., VINTERA J. *Chem. Listy* **50**, 1531 (1956).
8. BERGSTEIN A. *Silikáty* **3**, 161 (1969).
9. BLAŽEK A. *Silikáty* **1**, 158 (1957).
10. BLAŽEK A. *Hutnické Listy* **12**, 1096 (1957).
11. BLAŽEK A. *Hutnické Listy* **13**, 505 (1958).
12. BLAŽEK A., HALOUSEK J. *Hutnické Listy* **14**, 244 (1959).
13. BLAŽEK A., HALOUSEK J. *Automatic Balance*, Czechoslov. Pat. No. 92079 (1957).
14. CARPENTER H. C. H., KEELING B. F. E. *J. Iron Steel Inst. London* **65**, 244 (1964).
15. CLAISSE F., EAST F., ABESQUE F. *Use of Thermobalance in Analytical Chemistry.* Department of Mines, Province of Quebec, 1954.
16. COATS A. W., REDFERN J. P. *Analyst* **88**, 906 (1963).
17. ČERNÝ O. *Silikáty* **6**, 81 (1962).
18. DUVAL C. *Inorganic Thermogravimetric Analysis*, 2nd Ed. Elsevier, London, 1963.
19. DUVAL C. *Chim. Anal. (Paris)* **44**, 191 (1962).
20. DUVAL C. *Mikrochim. Acta* 705 (1958).
21. ELIÁŠ M., ŠŤOVÍK M., ZAHRADNÍK L. *Chemické rozbory nerostných surovin* No. 12. *Diferenčně termická analýza.* ČSAV, Prague, 1957.
22. FORMÁNEK Z., BAUER J. *Silikáty,* **1**, 164 (1967).
23. FOLDVARI-VOGL M. *Acta Geol. Acad. Sci. Hung.* **5**, 1 (1958).
24. FORMÁNEK Z., DYKAST J. *Silikáty* **6**, 119 (1962).
25. GARN P. D. *Thermoanalytical Methods of Investigation*, Academic Press, New York, 1964.
26. GARN P. D. *Anal. Chem.* **33**, 1247 (1961).
27. GOROCH G. L. *Diferentsialno termicheskii i vesovyi analiz*, Izd. Negra, Moscow 1964.
28. HARMELIN M. *La Thermoanalyse*, Presses Universitaires de France, Paris, 1968.
29. HAUPTMAN Z., STRNAD V., ŠKRAMOVSKÝ S. *Sborník pražské university universitě moskevské*, 1953, p. 299.
30. HONDA K. *Sci. Repts. Tohoku Imp. Univ.* **4**, 97 (1915).
31. IMRIŠ P. *Silikáty* **6**, 91 (1962).
32. *J. Thermal Analysis* **1**, (1969). Heyden, London; Akademiai Kiadó, Budapest.
33. KEATTCH C. J. *Talanta* **11**, 543 (1964).
34. KEATTCH C. J. *An Introduction to Thermogravimetry*, Heyden, London, 1969.

35. KISSINGER H. R. *J. Res. Nat. Bur. Stds.* **57,** 217 (1956).
36. KUZNIAROWA A. L. *Termogramy mineralov ilastych,* Wydanie geolog., Warsaw 1967.
37. LE CHATELIER *Compt. Rend.* **102,** 1243 (1886).
38. LE CHATELIER *Bull. Soc. Franc. Mineral. Crist.* **10,** 204 (1887).
39. LE CHATELIER *Compt. Rend.* **104,** 1443 (1885).
40. LE CHATELIER *Compt. Rend.* **104,** 1557 (1887).
41. LE CHATELIER *Z. Physik. Chem.* **1,** 396 (1887).
42. LE CHATELIER *Rev. Met. (Paris)* **1,** 134 (1904).
43. LUKASZEWSKI G. M., REDFERN J. P. *Lab. Pract.* **10,** 552 (1961).
44. MACKENZIE R. C. *The Differential Thermal Investigation of Clays,* Mineralogical Society, London 1957.
44a. MACKENZIE R. C., editor, *Differential Thermal Analysis,* Vol. 1. Academic Press, London (1970).
45. MACKENZIE R. C. *Differential Thermal Analysis Data Index "Scifax".* Cleaver-Hume Press, London, 1962.
46. MACKENZIE R. C., FARQUHARSON K. R. *Proc. 19th Session Intern. Geolog. Congress Algiers 1952,* No. 18, 183—200 (1953).
47. McADIE H. G. *Proc. First Toronto Symposium on Thermal Analysis,* Toronto 1965.
48. MIELEN R. C., SCHIELTZ N. C., KING M. E. *Proc. Nat. Conf. Clays and Clay Minerals, Missouri 1953. Natl. Acad. Sci., Natl. Res. Council Publ.* **327,** Washington 1954, p. 285.
49. MURPHY C. B. in *Treatise on Analytical Chemistry,* eds. I. M. Kolthoff and P. E. Elving, Vol. 8. Interscience, New York, 1968.
50. MURPHY C. B. *Anal. Chem.* **36,** 374 R (1964); **38,** 443 R (1966).
51. MURPHY C. B. *Anal. Chem.* **40,** 480 R (1968).
52. MURPHY C. B. *Anal. Chem.* **32,** 168 R (1960).
53. MURPHY C. B. *Anal. Chem.* **34,** 298 R (1962).
54. NERNST W., RIESENFELD E. H. *Ber.* **36,** 2086 (1903).
55. NEWKIRK A. E. *Anal. Chem.* **32,** 1558 (1960).
56. NUTTING P. G. *Standard Dehydration Curves, Proof Paper U.S. Geol. Surv.* No. 197 E, 1943.
57. PALEI P. N., SENTYURIN I. G., SKLYARENKO I. S. *Zh. Analit Khim.* **12,** 318 (1957).
58. PETRS H., WIEDEMANN H. G. *Z. Anorg. Allgem. Chem.* **300,** 142 (1959).
59. PILOYAN G. A. *Vvedenie v teoriyu termicheskogo analiza,* Izd. Nauka, Moscow, 1964.
60. PROCHÁZKA S. *Sborník prací z technologie silikátů,* SNTL, Prague, 1954.
61. REDFERN J. P. *Thermal Analysis, Proc. First Intern. Cong. Thermal Analysis,* Aberdeen Macmillan, London, 1965.
62. REDFERN J. P. *Thermal Analysis Review,* Vols. 1–6 (1964–1969).
63. ROBERTS-AUSTEN W. C. *Proc. Roy. Soc.* **49,** 347 (1891); *Proc. Inst. Mech. Engrs. (London)* **1,** 35 (1898).
64. ROBREDO J. *L'Analyse thermique différentielle en Verrerie,* Commision internationale du Verre, Paris, 1957.
65. ROSICKÝ J. *Silikáty* **12,** 295 (1968).
66. SALADIN E. *Iron and Steel Metallurgy and Metallography* **7,** 237 (1904).
67. SIMONS E. L., NEWKIRK A. E., ALIFERIS I. *Anal. Chem.* **29,** 48 (1957).
68. SLADE P. E., JENKINS L. T. *Techniques and Methods of Polymer Evaluation,* Dekker, New York, 1966.
69. SMOTHERS W. J., YAO CHIANG M. S. *Handbook of Differential Thermal Analysis,* Chemical Publishing Co., New York, 1966.

70. SOKOL L. *Silikáty* **1**, 177 (1957).
71. ŠATAVA J., VYTASIL V. *Silikáty* **1**, 185 (1957).
72. ŠATAVA V. *Silikáty* **1**, 188 (1957).
73. ŠATAVA V. *Silikáty* **1**, 204 (1967).
74. ŠATAVA V., TROUSIL Z. *Silikáty* **4**, 272 (1960).
75. ŠAUMAN Z. *Silikáty* **1**, 181 (1957).
76. ŠIŠKA V., PROKS I. *Silikáty* **5**, 142 (1961).
77. ŠKRAMOVSKÝ S. *Silikáty* **1**, 74 (1957).
77a. ŠKRAMOVSKÝ S. *Silikáty* **3**, 74 (1959).
78. ŠKRAMOVSKÝ S. *Chem. Listy* **26**, 521 (1932); *Collection Czech. Chem. Commun.* **5**, 6 (1933).
79. ŠKRAMOVSKÝ S., FORSTER R., HUTTIG G. F. *Z. Physik. Chem. B*, **25**, 1 (1934).
80. ŠPLÍCHAL J., ŠKRAMOVSKÝ S., GOLL J. *Chem. Obzor* **12**, 18 (1937); *Collection Czech. Chem. Commun.* **9**, 302 (1937).
81. ŠTÍMEL J., GÖMÖRY I., ČECH K. *Czechoslovak Pat.* 105 848 (1962).
82. *Talanta* **16**, 1227 (1969).
83. URBAIN G., BOULANGER C. *Compt. Rend.* **154**, 347 (1912).
84. VANIŠ V. *Silikáty* **4**, 266 (1960).
85. VOLDÁN V. *Silikáty* **1**, 125 (1957).
86. WENDLANDT W. W. *Thermal Methods of Analysis*, Interscience, New York, 1964
87. WIEDEMANN H. G. *Z. Anorg. Allgem. Chem.* **306**, 84 (1960).
88. WIEDEMANN H. G. *Z. Anal. Chem.* **231**, 35 (1967).
89. ZEDEK S., HYNEK S., BARTÁKOVÁ Z. *Silikáty* **1**, 141 (1957).
90. *Trudy pervogo soveshchaniya po termografii*, Izd. Akad. Nauk SSSR, Moscow, 1953.
91. ZÝKA J. *Analytická příručka*, SNTL, Prague, 1967 (Chapter *Termická analysa*).
92. 2000 *Differential Thermal Analysis Curves and Indices*, Sadtler Res. Lab., Philadelphia.
93. SCHULZE D. *Differentialthermoanalyse*, VEB Verlag der Wissenschaften, Berlin, 1969.
94. RAMACHANDRAN V. S. *Application of DTA in Cement Chemistry*, Chemical Publishing Co., New York, 1968.
95. MCADIE H. G. *Proc. Third Toronto Symposium Thermal Analysis*, Canadian Institute of Chemists, Toronto, 1969.
96. MURPHY C. B. *Anal. Chem.* **42**, 268 (1970).
97. SCHWENKER R. F. JR., GARN P. D. *Thermal Analysis, Proc. Second Intern. Conf. Thermal Analysis*, Worcester, U.S.A., Academic Press, New York, 1969.
98. HAINES P. J. *J. Chem. Educ.* **6**, 171 (1969).
99. MONTAMAT M. *Ind. Chim. (Paris)* **55**, 325 (1968).
100. MURPHY C. B. *J. Chem. Educ.* **46**, 721 (1969).
101. PAULIK J., PAULIK F., ERDEY L. *Hung. Sci. Instr.* No. 9 (1968).
102. TJUCHIYA R. *Tagaku To Kogyo (Tokyo)* **21**, 183 (1968).
103. BERG L. G. *Zavodsk. Lab.* **33**, 1286 (1967).

Chapter 2.

THERMOGRAVIMETRY

2.1. SHAPE OF THE THERMOGRAVIMETRIC CURVE, AND METHOD OF EVALUATION

Thermogravimetry is the method in which the weight of the sample is continuously recorded during heating or cooling. The thermogravimetric curve thus obtained, expressing the dependence of the weight change on temperature, gives information on the sample composition, its thermal stability, its thermal decomposition, and on the products formed on heating. The relationship between the rate of weight change and the temperature or time is expressed by the derivative thermogravimetric curve.

In order to understand the principle of the method and importance of both types of relationship, consider the analysis of the dehydration curves of $CuSO_4 . 5 H_2O$, shown in Fig. 2.1. The sample is heated in air at a linear rate of approximately 4 °C/min. It will be shown later that a change in heating rate may influence the course of the reaction and hence the shape of the curve. In order to make the explanation clearer, both curves in Fig. 2.1 are idealized to a certain degree.

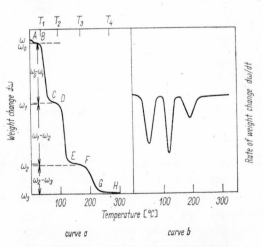

curve a   curve b

Fig. 2.1 Thermal analysis curves of a sample of $CuSO_4 . 5 H_2O$ (a) thermogravimetric (b) derivative thermogravimetric curve

19

Curve a represents the thermogravimetric curve, curve b the derivative curve. Consider first curve a: between points A and B no change in weight takes place, i.e. the sample is stable. At point B, dehydration takes place, is shown by a loss in weight. This loss ends at point C where the first which two molecules of water have been lost from the $CuSO_4 . 5 H_2O$. Between points C and D the sample is stable again, and at point D further dehydration takes place. The loss in weight between D and E indicates the loss of a further two molecules of water. The section between points E and F represents a stable compound which loses its last water molecule in the interval F to G. The interval G to H represents an anhydrous compound stable in the given temperature interval. Quantitative determination of weight changes is possible by measurement of the distance on the weight axis between corresponding horizontal levels. It is important to realize that the evaluation of the curve is to a certain extent subjective. The weight axis may be defined in various ways: (a) as a weight scale, (b) as a percentage of the total weight loss, (c) as an expression indicating the fraction decomposed, (d) as molecular units of weight. Generally, in thermogravimetry a curve of the decrease or increase in weight as a function of temperature is obtained, as can be seen from curve a. From this single curve, a series of important results can be obtained under suitable conditions. At the beginning of the curve a small loss in weight ($w - w_0$) is seen, corresponding to the desorption of the wash-liquid. If this loss in weight takes place at about 100 °C it is usually attributed to loss of adsorbed atmospheric moisture and a correction should be applied for it. The given substance is stable up to temperature T_1 at which it undergoes decomposition corresponding to a total weight loss of $w_0 - w_1$. In this case, the first dehydration step has taken place. The percentage decrease given by $[(w_0 - w_1)/w_0] \times 100$ often serves to identify the reaction which has taken place. The reaction ends at temperature T_2 at which another stable phase is formed which in turn dehydrates up to temperature T_3, and the new phase dehydrates up to temperature T_4, at which anhydrous sulphate is formed. It will be shown later that, analogous to other analytical methods, a number of variables influence reproducibility and shape of the TG curves. The main sources of error in this method are inaccurate measurement of temperatures, aerodynamic effects, effect of the temperature programme, and the influence of the furnace atmosphere and the heat of the reaction studied.

In a normal thermogravimetric curve, horizontal parts, which are characterized by a constant weight, can be interpreted, as can those parts where the slope giving the rate of weight change is steepest. The shape of the curves and the validity of an interpretation depend on the constancy of

experimental conditions. Often great importance is attached to accurate determination of the temperature at which the thermogravimetric curve begins
to deviate from the horizontal and which is often taken as the temperature
at which the reaction begins. In fact, however, this temperature is of
empirical value only, and depends on certain experimental conditions
connected with the apparatus used. From the thermogravimetric curve,
temperature intervals of thermal stability, and regions of formation and
existence of intermediates and the end-product, can be obtained, as is
evident from Fig. 2.1. The curve is also suitable for stoichiometric
calculations. However, the results obtained for the temperature of
single steps are somewhat empirical.

Curve b in Fig. 2.1 shows a derivative thermogravimetric DTG curve
and the peaks correspond to the single dehydration reactions described
above. Thermogravimetric curves normally express the dependence of the
sample weight (w) on temperature (T) or time (t)

$$w = f(T, \text{ or } t) \qquad (2.1)$$

whereas a derivative thermogravimetric curve shows the change in weight
with time (dw/dt) as a function of temperature or time,

$$dw/dt = f(T, \text{ or } t) \qquad (2.2)$$

Thus, this curve has a number of peaks instead of steps. In these curves the
area under the peaks (the integral) is proportional to the total change in
weight. The derivative curve may be obtained either by manual differentiation of the normal thermogravimetric curve, or by suitable instrumentation
described later in the appropriate chapter. DTG curves have certain
advantages over normal TG curves. They display a certain similarity to
the DTA curves, which allows, to a certain extent, comparisons to be
made [75, 108, 109].

The importance of the thermogravimetric method is due to the large
amount of information obtainable from a single measurement. Nev-
ertheless, for a true interpretation of thermogravimetric curves other
complementary methods also have to be applied, e.g. X-ray analysis,
evolved gas analysis, and differential thermal analysis.

2.2. FACTORS AFFECTING THERMOGRAVIMETRIC MEASUREMENTS

In the period 1950—1967 numerous designs of apparatus for thermogravimetry were developed, often differing very much in construction.

Consequently papers appeared in the literature, containing results which differed, not only with respect to the course and the shape of the thermo-gravimetric curve, but also with respect to the temperatures at the beginning and the end of the reaction. This occurred even in cases where the same compound was analysed. These discrepancies led to careful analysis of their causes, and to an exact study of conditions and their resulting effects. It was found that a number of factors must be controlled carefully if useful and reproducible thermogravimetric curves are to be obtained. These factors include the set-up of the measuring device (furnace and sample carrier), the quality and physical properties of the sample, the type of working atmosphere and the method of heating. The factors affecting the results of thermogravimetry can be divided into the following groups, which will be discussed in detail.

I. Effects arising from the properties of the construction material employed and the general nature of the apparatus.
  (a) Reactions of the sample with the crucible, and of the reaction products with parts of the apparatus.
  (b) Sublimation and condensation of the reaction products.
  (c) Shape, size, and material of the crucible.
II. Effects resulting from construction and method of use.
  (a) Heating rate and heat transfer.
  (b) Rate of recording the curve.
  (c) The composition of the atmosphere in the reaction chamber (furnace).
  (d) Nature of the heating, and the effects of buoyancy.
  (e) Sensitivity of the balance and the recording system.
  (f) Method of temperature measurement.
III. Effects of physical and chemical properties of the samples.
  (a) Amount of the sample.
  (b) Particle size.
  (c) Heat of reaction, and thermal conductivity of the sample.
  (d) Nature of sample and type of changes taking place.

Effects arising from the nature of the construction material employed

(a) REACTION WITH THE APPARATUS. Suitable construction of the apparatus enables the majority of the errors originating from the factors mentioned to be avoided. Such factors are mainly the electrostatic effects on the weighing mechanism, the effect of the furnace construction, and inac-curacies in the recording and weighing mechanism. These can be allowed for in the construction of the apparatus. Thus, for example, non-inductive winding of the furnace prevents the creation of a magnetic field, placing of the furnace above the balance reduces the possibility of the formation of convection currents, radiation shields diminish radiation in the direction of the weighing mechanism. Further effects included in the first group should be considered in every analysis, regardless of the construction of the balance. These include the possibility of reaction of the sample with the crucible, which may lead to completely erroneous results and conclusions. Newkirk and co-workers [72] found, for example, that the decomposition temperature of sodium carbonate is lower in a quartz or porcelain crucible than in a platinum one. A reaction between the sample and the quartz or porcelain crucible took place at approximately 500 °C, forming sodium silicate and carbonate, which distorted the TG curve. This effect can be still more pronounced during the analysis of polymers. It is known, for example, that polytetrafluorethylene and some other polymers react under certain conditions with porcelain, glass, or quartz crucibles, forming volatile silicate compounds [24]. Similarly, platinum is also attacked by polymers containing phosphorus, halogens or sulphur. The author found a similar effect when heating basic chromium slags in corundum crucibles, and when platinum crucibles were attacked during the analyses of sulphite lyes and phosphoric acid amides. However, the effect of the crucible is not restricted to reaction with the sample. The effect may be due to porosity of the crucible, or the catalytic effect of the crucible material, or the previous history of the crucible. Hence, increased attention should be paid to the possibility of the interaction of the crucible material with the sample, and materials used for the crucible and sample carrier should be sufficiently inert towards the investigated substance, decomposition products, and gases in the furnace. The working atmosphere may have a negative effect, not only on the metallic parts of the apparatus mentioned above, but also on the thermocouples and on various parts of the weighing mechanism. Thus, increased demands on the construction of the apparatus and on the material used for the construction of the weighing mechanism are made, and the importance of the calibration of the balance and the system for measuring the temperature (to be discussed in a later chapter) becomes apparent. The corrosive effect of gaseous products should be considered in overall design of the thermobalance. The problem is usually solved by using special materials or by employing a special system for flushing the weighing mechanism, or by enclosing it in an atmosphere of inert gas [8, 21].

(b) SUBLIMATION AND CONDENSATION. Another important factor is the condensation of volatile products on cooler parts of the weighing system. This effect is described in some older papers [24, 90] and it was observed by the author during the thermal analysis of arsenopyrite [7], when As_2O_3 condensed on the cooler parts of the suspension, and evaporated when the temperature was further increased. The TG curves were thus quite unreproducible. The condensation is affected by the rate of gas flow, as the volatile products can be carried to various parts of the suspension. The best method of deciding whether such effects have taken place during the experiment is control weighing of the crucible, sample, and the suspension before and after the experiment. Unfortunately, the construction of the majority of thermobalances does not permit this. The most efficient way of eliminating this source of error is to place a ceramic or heat-resistant tube around the crucible carrier. This tube prevents condensation on the crucible support. The author eliminated this source of error in the case of a thermobalance used for work in a defined gaseous medium [8], by placing the furnace above the balance, using a ceramic protecting tube around the suspension (which was provided with several radiation shields), and by having a layer of inert gas in the space between the protecting tube and the suspension.

(c) THE CRUCIBLE. The size and shape of the crucible can also appreciably influence the course of the thermogravimetric curve. The crucibles described in the literature differ widely in shape, size and construction material. Crucibles will be described later in Chapter 5. At this point it is sufficient to note that the shape and size of the crucible can affect the TG curves. If the volatility of a substance is followed under isothermal conditions, the rate of volatilization depends on the shape of the crucible, as this determines the surface area of the sample. The size of the crucible and the amount of the sample play a major role as they affect the method of heating and also the rate of diffusion of gases liberated from the sample. In view of this, it appears that the best results are obtained with crucibles in the shape of a small shallow dish permitting homogeneous heating of the sample over a relatively large surface area. It is important to take into account other properties of the investigated substance such as decrepitation, or foaming, during heating. If any of these occurs the crucible described is unsuitable, and cylindrical or conical or closed crucibles should be used. The shallow crucible is also unsuitable in cases where a gas stream is used, as aerodynamic effects cause complications. The effect of the size and the shape of the crucible was investigated by Garn and Kessler [37, 38] mainly with respect to the weight and thermal conductivity. They found that in the case of a heavy crucible the decomposition of the sample takes place in a smaller temperature interval (as in the case of $PbCO_3$). It is desirable that the size and shape of the crucible are such that rapid heat transfer and homogenous heating occur.

This problem was investigated by several authors [58, 61, 72, 76] who came to the conclusion that the construction of the crucible can influence the course of the decomposition curves. The effect of the material is negligible if Pt, Au or Al crucibles are used. The crucible should be symmetrical, small, and of a suitable shape to facilitate gas flow in the furnace. The use of a crucible closed with a piston or a lid may appreciably affect the course of decomposition reactions, owing to the pressure of the liberated gases, and changes in partial pressure of gases in the furnace atmosphere. The length of the horizontal part of the TG curve is increased when closed crucibles are used, and therefore their use is not usually recommended, except in cases of decrepitation or substances which creep up the walls of the crucible. The question of the shape and size of the crucible is important in relation to the amount of sample used and heat transfer during measurement of kinetic data (mainly by the dynamic method) when a knowledge of the true temperature of the sample, which must be uniformly heated, is of prime importance in obtaining meaningful results.

Effects resulting from construction and method of use

(a) EFFECT OF HEATING RATE AND HEAT TRANSFER. Thermogravimetry requires heating and weighing of the sample simultaneously. This means there must be no contact between the sample and the furnace wall. Thus the problem of heating the sample is primarily one of heat transfer. This is controlled by a number of factors, e.g. the properties and size and nature of the sample, and the enthalpy change of any reactions undergone by the sample. A temperature gradient is formed between the sample and the furnace wall, which introduces errors into measurement of the sample temperature and determination of the range of temperature reactions being investigated. A temperature gradient may also be formed inside the sample. The temperature difference between the furnace wall and the sample is affected primarily by sample properties, the enthalpy change of any reactions it undergoes, and the rate of temperature increase. This difference, which is substantial in the formation of a DTA curve, complicates the measurement of temperature of the investigated sample in this case, and may lead to serious errors. This is true mainly of those types of thermobalance where the temperature is measured with a thermocouple placed in the neighbourhood of the sample or the furnace wall, but not in the sample itself. The error is also connected with the temperature programme, which is usually linear and creates a certain temperature difference between the furnace and the sample. When the temperature is measured with a thermocouple placed outside the sample the values measured are usually higher when the sample is continuously heated than when it is heated in a series of steps. This is due to the tem-

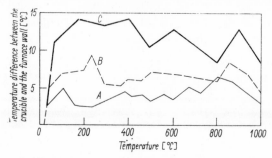

Fig. 2.2 Effect of heating rate on the temperature difference between sample holder and furnace. Rates of heating: A — 150 °C/hr; B — 300 °C/hr; C — 600 °C/hr. (Reprinted from [71] by permission of the copyright holder)

perature gradient between the sample and the furnace, so that on continous heating the sample temperature lags behind that of the furnace. This problem was investigated by Newkirk [71] who used a Chevenard thermo-balance and measured the temperature difference between thermocouples placed inside the crucible and near the furnace wall.

The results of the measurements are shown in Fig. 2.2. It is seen that the observed temperature difference varies between 3 and 14 °C and is approximately proportional to the temperature increase. Hence, the error in the measured temperature will be small at low rates of heating. Thus kinetic data calculated from thermogravimetric curves obtained at relatively high rates of heating, e.g. 10 °C min, which is often used in thermogravimetry, will not be reliable. The temperature difference is also affected by the enthalpy changes of any reactions the sample undergoes. This effect depends on a number

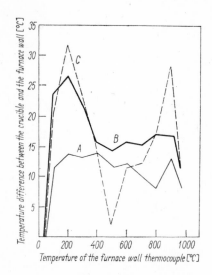

Fig 2.3 Effect of sample on the temperature diffence between crucible and furnace. Decomposition of $CaC_2O_4 \cdot H_2O$ at 600 °C/hr heating rate.
A — crucible alone; B — 9.2 g of sample; C — 9.6 g of sample. (Reprinted from [71] by permission of the copyright holder)

of factors, but mainly on whether the reaction is endo- or exothermic, and also on whether the heat capacity of the sample changes during the experiment. This effect can be seen in Fig. 2.3 which shows New-kirk's results for a $CaC_2O_4 . H_2O$ sample.

Here the difference between the temperatures of the sample and the furnace wall was measured with various amounts of the sample and at a relatively high rate of heating (10°/min). The crucible temperature was lower than the furnace temperature by 10–14 °C up to 1000 °C. With a sample weight of 0.2 g, the endothermic effect of dehydration caused a difference of 25 °C at 200 °C; the exothermic loss of CO almost compensates for this difference, but the difference increases again owing to the endothermic loss of CO_2. With larger amounts of sample both these effects increase. If the reaction is endothermic the temperature difference is increased and if it is exothermic the difference decreases. As a high rate of heating increases the temperature difference, the use of higher heating rates has become general practice in differential thermal analysis. The effect of heating rate on the course of the thermogravimetric curve was investigated by several authors [22, 27, 61, 67, 71, 80, 83] who found that generally speaking the lower the rate of heating the lower the temperature of decomposition and the lower the temperature at the end of the reaction. A change in the rate of heating may separate consecutive reactions; thus if during rapid heating an inflexion point appears on the curve, on slower heating it may change to a more or less well-formed plateau. This change is important in the identification of intermediates and has led to varying interpretations of results. For example, during the decomposition of $NiSO_4 . 7 H_2O$ at a heating rate of 2.5 °C/min [23] only the plateau of the monohydrate was observed, whereas at a 0.6 °C/min heating rate the existence of hexa-, tetra-, di-, and monohydrates was detected [35]. This shows that in analysing a sample containing a large amount of bound water it is necessary

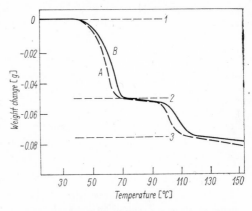

Fig. 2.4 Effect of heating rate on the shape of the thermogravimetric curve of a sample of $Na_2B_4O_7 . 10 H_2O$.
Level 1: $Na_2B_4O_7 . 10 H_2O$; Level 2: $Na_2B_4O_7 . 5 H_2O$; Level 3; $Na_2B_4O_7 . 2.5 H_2O$
Curve A: heating rate 1 °C/min;
Curve B: heating rate 5 °C/min.

to use low rates of heating. A change in heating rate during an experiment can also lead to the formation of inflexions on TG curves [59]. This effect should be related to factors which are affected by the rate of heating, such as the composition of the furnace atmosphere, which depends on the rate of formation of the reaction products. Thus, for example, during the thermal treatment of pyrites the author showed more $FeSO_4$ was formed on slower heating [7]. The rate of temperature increase and measurement of the true temperature of the sample are of basic importance in the determination of kinetic data from thermogravimetric curves. The effect of the rate of heating on the analysis of borax is shown in Fig. 2.4.

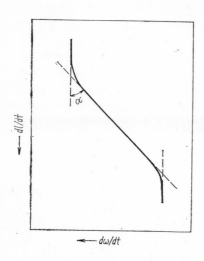

Fig. 2.5 Effect of chart-speed on the resolution factor R. Optimum case: $\alpha = 45\ °C$. (Reprinted from [59] by permission of the copyright holder)

(b) EFFECT OF THE RATE OF RECORDING OF THE TG CURVE. The rate of recording of the TG curve may have an appreciable effect on its clarity and shape. Lukaszewski and Redfern [59] introduced the concept of a resolution factor R, given by

$$R = \tan \alpha = r_2/r_1 \qquad (2.3)$$

where r_1 is the rate of recording in cm/min and r_2 is the rate of weight change expressed on the recording as cm/min. The slope of the curve is dependent both on the recording rate and on the scale used for the weight change. For accurate location of the mid-point of the inflected curve it is desirable that $\alpha = 45°$ (for $r_1 = 1$ cm/min). For the usual heating rates $(1-6\ °C/min)$ the chart-speed should be $15-30$ cm/hour (see Fig. 2.5). A more rapid movement may lead to an undesirable distortion of the curve. Some instruments are provided with X, Y-recorders and then the slope of the curve often cannot be adjusted to the optimum for varied heating rates, and it seems more advantageous to have a free choice of chart speed and rate of temperature increase.

(c) COMPOSITION OF THE ATMOSPHERE IN THE REACTION SPACE. The atmosphere in the reaction space of the furnace is one of the major factors contributing to variations in results obtained by different authors.

If the sample is dried or decomposed in air, the atmosphere in the neighbourhood of the crucible is continually changing owing to liberation of gaseous products or reaction of the sample with the original atmosphere. In the first case dissociation takes place and the dissociation temperature depends on the partial pressure of the gaseous components in the vicinity of the sample, because dissociation takes place only when the dissociation pressure exceeds the partial pressure of the ambient gas. Closing the crucible can affect the partial pressure of the gas inside it and thus the dissociation temperature. An interesting application of the controlled atmosphere in thermogravimetry is the separation of curves, for example; the curves of calcite and dolomite are superimposed in air but not in a carbon dioxide atmosphere [40]. As the concentration of a gaseous product in the vicinity of the sample increases, the reaction rate decreases. Hence, the composition of

the furnace atmosphere may appreciably affect the TG curve, even when no direct re- action occurs between the sample and some of the at- mospheric components. When programmed heating is applied the gas concentration in the vicinity of the sample changes as a result of buoyancy effects, creating atmospheric condit- ions which are hard to repro- duce. The heated sample is not in equilibrium with the ambient atmosphere at the moment of decomposition, but it approaches equilibrium as a plateau is formed on the TG curve. Slow heating ensu- res constancy of the atmosphe-

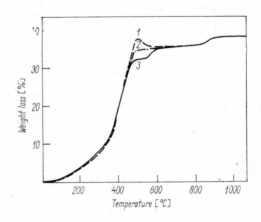

Fig. 2.6 Effect of the furnace atmosphere on the TG curves of $MnCO_3$ *1* — Weight 2.5 g; *2* — Weight 2.0 g; *3* — Weight 0.1 g.
(Reprinted from [9] by permission of the copyright holder)

ric composition, causing the decomposition reaction to take place within a narrower temperature interval, as well as giving a uniform temperature inside the sample. Thermogravimetry is often carried out in a dynamic atmosphere of the gas(es) produced by the reactions of the sample. Construction of the furnace parts may permit dissipation of liberated gases to the surroundings, giving rise to different partial pressures of gaseous components and thus affecting the course of the curve. This occurs, for example, in the decomposition of carbonates. The effect on the TG curve of $MnCO_3$, when this decomposes in its self-generated atmosphere. is shown in Fig. 2.6 [6, 9].

The concentration of CO_2 in the furnace atmosphere, created by the decomposition of $MnCO_3$, may affect the oxidation of reactive MnO formed during the decomposition. However, the effect of the furnace atmosphere may be more complicated than simply affecting decomposition or oxidation. For example, during the thermal treatment of FeS_2 in air [7], decomposition to a mixture of Fe_2O_3 and $FeSO_4$ takes place within the temperature range 390—500 °C. The sulphate formed (the amount of which depends on the amount of sample, the access of oxygen to it and the heating rate) decomposes in the range 520—560 °C (see Fig. 2.7). With a 0.1 g sample of FeS_2 the decomposition curves of FeS_2 and $FeSO_4$ are distinctly separated, but with higher weights both reactions take place simultaneously, the final product being Fe_3O_4 and not Fe_2O_3 as in the first case. A more complicated situation arises in the case of a mixture of $MnCO_3$ and FeS_2 [10] (see also Fig. 2.7). At about 400 °C, decomposition of both $MnCO_3$ and FeS_2 occurs, $MnSO_4$ and $FeSO_4$ being formed. The iron sulphate decomposes in the range 520–620 °C and the mangan-

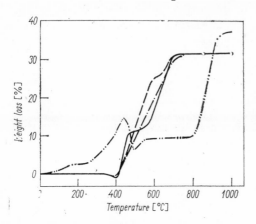

Fig. 2.7 Effect of the furnace atmosphere on the TG curves of FeS_2 and a mixture of $MnCO_3$ and FeS_2
—— FeS_2 0.1 g; — — — FeS_2 0.5 g;
—.— FeS_2 1.0 g; —..— $MnCO_3 + FeS_2$
(1 : 1) 0.1 g. (Reprinted from [10] by permission of the copyright holder)

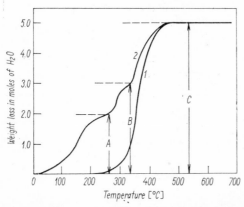

Fig. 2.8 Effect of furnace atmosphere on the course of dehydration of $Cd_5H_2(PO_4)_4(H_2O)_4$ Curve 1: open crucible; curve 2: closed crucible—self generated atmosphere. Level A: $Cd_5H_2(PO_4)_4(H_2O)_2$— loss of 1.9 molecules of H_2O; Level B: $Cd_5H_2(PO_4)_4(H_2O)$—loss of 3.0 molecules of H_2O; Level C: $Cd_3(PO_4)_2 + Cd_2P_2O_4$—loss of 5.05 molecules of H_2O. (Reprinted from [81] by permission of the copyright holder)

ese sulphate in the range 780–920 °C. The weight of the sample affects the composition of the furnace atmosphere, which in turn affects the course of both sulphation reactions. Another possible complication, indicated by Guiochon [42], is the dissolution of atmospheric components in liquids or solids formed during the reactions. This effect cannot be eliminated easily and it usually remains unknown. It was demonstrated that with a sample of NH_4NO_3 a substantial part of the HNO_3 could not be eliminated even after two hours' treatment at 200 °C. An example of the effect of atmosphere is provided by the dehydration of $Cd_5H_2(PO_4)_4 \cdot (H_2O)_4$ [81] (see Fig. 2.8). In an open crucible dehydration takes place continously from 250 °C, whereas when a water-saturated atmosphere is used the waves for mono- and dihydrate are separated.

Great attention has been paid to the effect of the furnace atmosphere not only on hydrates, where the water vapour pressure affects stability and the course of dehydration, but also carbonates, organic substances, etc. [27, 36]. Vallet [104, 105] studied the effect of water vapour pressure on the decomposition of $CuSO_4 \cdot 5H_2O$, and of CO_2 pressure on the decomposition of $CaCO_3$, and also their effect on the equilibrium temperature. Garn and Kessler [39], who heated the samples in open and closed crucibles, studied the effect of the atmosphere on the course of the decomposition of $MnCO_3$, $PbCO_3$, $(NH_4)_2CO_3$ and $CoC_2O_4 \cdot 2H_2O$. Generally, it can be said that an increase in the concentration of a gaseous component liberated by the reaction leads to the suppression of the reaction. The composition of the furnace atmosphere can be affected by an external influence, e.g. by the introduction of gas of a certain composition, and at a given flow-rate. At every temperature there is a characteristic non-equilibrium dissociation pressure of the gases over the sample. If another gas is introduced into the furnace, decomposition takes place at the temperature at which the non-equilibrium dissociation pressure exceeds the partial pressure of the gas introduced.

When considering the effect of atmosphere on the course of the TG curves two factors have to be taken into account (bearing in mind the results required from the analysis): (a) whether a static or dynamic atmosphere has been used and (b) whether the sample was heated in a defined atmosphere (e.g. N_2, O_2, H_2), or in the medium of its own gaseous reaction products. If a self-generated atmopshere is used it is evident that when closed crucibles are used the TG curve undergoes a shift to higher temperature (for reasons given earlier). On the other hand, the influence of reactive atmospheres created by the introduction of a gas from an external source may affect not only the position of the observed change on the temperature axis, but may also change the reaction itself (e.g. oxidation, reduction). This is also true of the gas pressure in the furnace atmosphere. The rate of oxidation, for example, increases in various ways with an increase in oxygen. The rate and type of flow of the furnace atmosphere can affect the accuracy of the weighing, as will be shown later, but it can also change the composition of the atmosphere in close proximity to the sample (e.g. by removal of the reaction products), and thus affect the course of the TG curve.

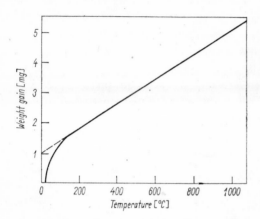

Fig. 2.9 Correction curve for apparent weight increase as a function of temperature, for the Chevenard thermobalance. (Reprinted from [84] by permission of the copyright holder)

(d) NATURE OF THE HEATING AND EFFECT OF BUOYANCY. In the preceding section it was stressed that the rate and type of gas flow in the reaction space of the furnace may appreciably affect the accuracy of weighing. According to Duval [27] this is one of the most important factors influencing results. Movement of the atmosphere in the reaction space occurs by convection as well as by the introduction of gas from an external source. Owing to heat-transfer effects the furnace walls are hotter than the sample, and as heat transfer is by convection at lower temperatures, the rising of gas along the furnace walls displaces gas down the centre of the furnace, thus causing an apparent increase in weight. This effect was studied in detail by Simons, Newkirk, and Aliferis [84], using a Chevenard thermobalance. Using platinum, molybdenum and porcelain crucibles they obtained the correction curve shown in Fig. 2.9. When a 200 mg crucible is used a total weight increase of 5 mg occurs at 1073 °C.

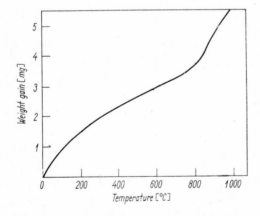

Fig. 2.10 Correction curve for convectional gas flow in the Chevenard thermobalance. (Reprinted from [68] by permission of the copyright holder)

The rate of weight increase is greatest in the range up to 200 °C, while in the range 200–1000 °C the increase is linear. A similar correction curve was constructed with the same apparatus by Mielenz, Schieltz, and King [68] who used a sample weighing 1.5 g. Their curve displays approximately the same general weight increase at 1000 °C, but in the 700–800 °C temperature interval it is non-linear (Fig. 2.10). Lukaszewski [60] also confirmed the existence of this phenomenon and that it was due to an appreciable temperature difference between the furnace wall and the crucible. Changes in temperature will cause changes in the density of the gaseous medium which surrounds the sample. The apparent weight increase can be expressed by the relationship

$$w = Vd(1 - 273/T) \qquad (2.4)$$

where w = apparent weight increase, V = volume of the sample, crucible, and carrier, d = density of air at 273 K, T = temperature in K [18]. Newkirk studied this effect in the temperature range 25–1000 °C, using a Chevenard thermobalance, and he found that in the case of hydrogen the weight increase was 0.1 mg, for air it was 1.4, and for argon 1.9 mg.

From the literature it is evident that the effects observed are caused by changes in density of the heated gas and by convection currents. The net effect is influenced predominantly by the construction of the working space, and the size and shape of the crucible and carrier. There are a number of ways of reducing or eliminating this effect, such as reduction in size of the carrier and crucible. However, as a certain amount of sample is required, and as other factors such as the requirement of a large surface area must be considered, this is not always possible. The same is true of the use of thin aerodynamic sample-carriers and crucibles [13, 60]. Duval [27] reduced the effect due to convection currents by suitable ventilation of the furnace, enabling gas to flow from the lower to the upper part of the furnace. By providing the upper furnace lid with openings of variable size, he succeeded in suppressing the effect quite appreciably when using a cylindrical crucible. In assessing the suitability of an analytical precipitate for thermogravimetric analysis, this method is satisfactory, consisting of a trial and error adjustment of ventilation of the furnace space to make the apparent weight increase negligible for a given crucible and a given amount of sample. The effect of ventilation on the correction curve is shown in Fig. 2.11 [71]. However, for a correction of this type the conditions are valid only for the given type of furnace, crucible, and carrier, and it is necessary to determine a correction curve in each case. Another method of reducing the effects is the use of a double furnace. This is often preferable as it can also eliminate the disturbance caused by the rate of flow of gas introduced into the furnace. If this method is used, suspensions and samples of approximately the same weight and geometric shape should be attached to both beams of the balance and put into two furnaces with identical working spaces in which both the gas introduced and the heating are the same. In this way buoyancy effects are eliminated, being identical in both furnaces, and only the sample weight change proper appears. This method requires special construction of the balance. It is also possible to use the technique (due to Keyser [54]) of differential weighing which will be described later. However, in many cases these methods are unsatisfactory and the TG curve can be better corrected mathematically on the basis of several control experiments carried out under conditions identical to those chosen for the analysis of the substance [20, 71, 63]. One method of minimizing the effects, which has been introduced recently in certain commercial thermobalances, is the miniaturization of the apparatus, mainly the furnace.

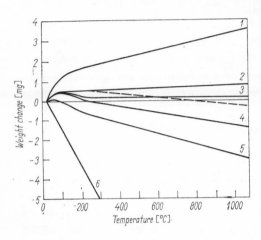

Fig. 2.11 Effect of furnace-lid opening on the 'apparent' weight increase or decrease. Chevenard thermobalance, rate of heating 300 °C/hr. porcelain crucible 4 g, dotted curve for two crucibles in double furnace. *1* — lid without opening; *2* — diameter of the opening 7 mm; *3* — 8 mm; *5* — 10 mm; *6* — 43 mm (without lid). (Reprinted from [71] by permission of the copyright holder)

A further effect on the TG curve arises from working with a controlled atmosphere, i.e. with gas introduced into the furnace at a certain rate from an external source. This effect depends primarily on the dimensions and the geometry of the working space and the gas flow-rate, and should be considered in each case. Newkirk, using a Chevenard balance, found an apparent weight increase which increased with the rate and the amount of gas introduced. This effect can be reduced by suitable construction of the working space and by preheating of the introduced gas. The most suitable method consists in placing the sample and the furnace above the balance, and filling the upper part of the furnace with glass wool or other heat-resistant material in which the gas is uniformly preheated and then flows at a constant rate towards the crucible and sample. Attempts

have also been made to study the buoyancy effects under conditions os
elevated or reduced pressure. However, it seems that under such conditionf
the situation is much more complicated and it will not be discussed further.
It is evident that the appearance of various fluctuations and apparent
losses or increases in weight (described above) should be the result of the
aerodynamic effects rather than any factor in the construction of the
weighing mechanism. For this reason these effects should be considered
separately in each case with respect to construction and arrangement of the
apparatus. This is illustrated by the results of Cahn and Peterson [14],
who systematically followed the effect of the diameter of the reaction tube
on the sensitivity and the stability of recording under conditions of
atmospheric pressure and a streaming atmosphere, using an electronic
balance of the Cahn RG type (see Chapter 5). They determined the
"noise" on the TG curve at various values of initial parameters and found
in all cases negligible noise (approx. 1–3 µg) at a reaction tube diameter
below 17 mm. The observed noise was maximum in the 150–650 °C
temperature interval.

The importance of composition of the atmosphere in solid–gas reactions is generally acknowledged today and great attention is devoted
to it in the development of commercial thermobalances. For example, the
Mettler thermobalance makes use of several types of furnace space. An
interesting method is used in the Perkin–Elmer TGS-1 thermobalance.
It uses a microbalance requiring samples weighing only milligrams,
and a miniature furnace in close proximity to the crucible. This
arrangement provides good contact between the atmosphere and sample, and stability of the zero line, and it would appear that the
problem of buoyancy is much less than in the case of the Chevenard-
type balance.

(e) EFFECT OF THE SENSITIVITY OF THE BALANCE AND THE RECORDING SYSTEM.
The question of the sensitivity of the weighing mechanism and the
recording system is closely related to the amount of sample used. The
weight of the sample may exert an appreciable influence on the course
of the reactions being followed. Often, in order to follow the reaction
under required conditions, as low a sample weight as possible is taken.
However, this places increased demands on the sensitivity and stability
of the whole apparatus. Great attention must always be devoted to
long-term stability of the thermobalance, periodic calibration of weight
ranges, quality and strength of the record, etc. These factors will introduce
errors into the thermogravimetric measurement mainly in prolonged experiments and at high sensitivities of measurement. These problems are

directly connected with the construction of the weighing and the recording system and they will not be discussed in greater detail at this point.

(f) METHODS OF TEMPERATURE MEASUREMENT. Methods and possible methods of temperature measurement both in thermogravimetry and in DTA will be discussed in a separate chapter. Here, only some basic conditions will be indicated, mainly the correct placing of thermocouples. Neglect of this can lead to serious errors. In the section on heat transfer to the sample it was shown that heating of the sample occurs by transfer of heat from the furnace walls to the crucible and further to the sample without direct contact. This transfer is affected by a number of factors, such as the heat conductivity of the atmosphere, heat conductivity and heat capacity of the crucible and sample. Examples were given earlier showing that a temperature gradient is created between the furnace wall and the sample, and considerable temperature differences are often observed. These differences may be affected by the heat of reactions occurring. Therefore, it is very important that the sample temperature is as homogeneous as possible and that the thermogravimetric curve be related to the temperature measured in the sample, and not near the furnace wall or in the furnace space outside the sample, as is often the case.

This is important not only in the determination of thermal stability of substances and of reactions they undergo, but also in the use of the thermogravimetric curve in calculating thermodynamic data, when an accurate knowledge of the true temperature of the sample is especially important.

Another important factor in temperature measurement is the fact that when a deflection-type balance is used the beam and thus the crucible move as the weight changes. In view of this it is important that the construction of the crucible carrier ensures the placing of the sample in a constant-temperature region of the furnace, which should be constructed so as to make this zone as large as possible. This problem is especially important when a simultaneous DTA is made, as the DTA curve may be appreciably distorted by movement of the crucible in the furnace. Thus in the case of combined measurements preference should be given to thermobalances working on the null-point principle, where the balance beam movement is limited to a minimum, ensuring a practically constant position of the sample in the furnace.

Effects of the physical and chemical properties of the sample

(a) AMOUNT OF SAMPLE. The most important sample properties are weight, particle size, type of packing in the crucible, and reactions which it undergoes during heating. However, these properties should be considered carefully, taking into account possible reactions with the crucible material and the possible changes of state. One of the most important parameters is its quantity. It is known that with a powdered sample the larger the initial sample weight, the larger also the temperature drop inside it, (as shown in Fig. 2.3). In consequence, the time required for the reaction to take place throughout the whole volume of the sample is prolonged. The heat of reaction thus causes a difference between the temperature of the sample and that of the furnace, depending on the magnitude of the heat of reaction, and thus will affect the temperature drop inside the sample as mentioned above. TG curves obtained under such conditions cannot be used for the determination of reaction kinetics and even their analytical evaluation requires careful consideration of experimental conditions. From isothermal kinetic studies it is known that the size and the geometry of the sample affects the values of the kinetic data; the lower the sample weight, the larger the values of the calculated rate constants of the reactions [4, 69]. With higher weights clarity of curves is usually worse and they are shifted to higher temperatures. This is one of the many reasons for discrepancies between the experimental results of various authors using different types of thermobalance, with different weights of sample. The sample weight can influence the shape and course of the TG curve not only as a result of the temperature gradient inside the sample, but also indirectly by affecting the furnace atmosphere, as was shown in the example of the $MnCO_3$ decomposition curves obtained with different sample weights (see Fig. 2.6).

(b) PARTICLE SIZE. Another important property of the sample, which might affect the course of the TG curve, is the particle size. It is known that the more finely ground a solid is the more reactive it is, and this directly affects the shape of the curve. In a series of investigations it was shown that the more finely divided the sample the lower the temperature at the beginning and end of the reaction, and that the reaction is accelerated [9, 66, 73, 82, 104, 105].

This effect is shown in Fig. 2.12; the TG curve of dialogite ($MnCO_3$) was obtained with the unground crystalline substance and with finely ground material [9]. In the case of the unground sample, decomposition takes place at 570 °C and the curve passes through a maximum corresponding to a temporary formation of MnO. At 830 °C a plateau is attained corresponding to Mn_2O_3 which changes to β-Mn_3O_4 at 930 °C. In the case of a finely ground sample the decomposition starts at 390 °C, and continues slowly up to 570 °C and then rapidly up to 690 °C, forming a plateau corresponding to Mn_2O_3. Particle size may affect the time of access of the furnace atmosphere to inner parts of the sample and thus prevent instantaneous oxidation of the decomposition product formed, in this case MnO. The course of the TG curves may also be affected by the history of the sample under investigation, e.g. minerals and voluminous precipitates. Therefore, it is very important, in order to compare thermogravimetric results, that details of the experimental conditions be accompanied by a detailed description of the sample. This is recommended by the standardization committee of ICTA, for presenting results [65]. In the case of minerals, the time since the sample was obtained from its source is also important as exposure to the air may cause a change in its chemical and physical properties, e.g. the loss of water, which leads to a change in thermal conductivity. The preliminary treatment and the history of the sample under analysis are very important in the case of moist precipitates where the amount of water retained not only reduces the thermal conductivity of the sample, but also affects the reaction atmosphere in the furnace (as water vapour is liberated). This should be borne in mind in the analysis of moist analytical precipitates by Duval's method. A typical example is given by Rynasiewicz and Flagg [81a]. There are a number of other sample properties which might influence the course of the TG curve, such as the catalytic effect of impurities, the effect of a change in crystal structure or other transformations, the effect of melting of the sample, the dissolution of gases in the sample, and the packing of the sample in the crucible. Some of these effects cannot be foreseen and they must always be considered with respect to the particular sample being analysed.

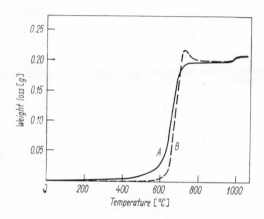

Fig. 2.12 Effect of the nature of the sample particles on the course of the TG curve.
A — Dialogite (from Biersdorf), finely ground.
B — The same sample coarsely crystalline.
Atmosphere: static air. Heating rate: 6 °C/min.
Weight 0.62 g.

(c) HEAT OF REACTION AND THERMAL CONDUCTIVITY OF THE SAMPLE. The effects of these two factors are interconnected. An exothermic reaction will raise the temperature of the sample and so reduce the temperature lag between furnace and sample; an endothermic reaction will similarly increase the temperature lag. The effect will be dependent on the thermal conductivity of the sample, however, since the reaction will be initiated in the hottest part of the sample, i.e. that nearest the furnace wall, and propagation will depend on the rate at which the heat evolution or absorption is transmitted to the rest of the sample. If heat is absorbed then there will be a delay before the rest of the sample reaches the critical reaction temperature and the apparent reaction temperature (as measured by the thermocouple in the furnace wall) will be higher than the true reaction temperature. In any case, the temperature lag will depend on the thermal conductivity of the sample, if the sample is more massive than the sample holder [72].

(d) NATURE OF THE SAMPLE AND TYPE OF CHANGES TAKING PLACE. A change in the specific heat of the sample (as for example at the Curie point of magnetic material, order-disorder transitions, etc.) will necessarily affect the thermal conductivity. Magnetic materials will interact with the field of the furnace winding, and cause difficulty. Many polymers, if prepared in powder form, are prone to become electrostatically charged, and difficult or even impossible to get into a sample holder. Electrostatic charges will naturally induce charges on the balance parts and cause difficulty. The production of gases in the thermal decomposition of the sample will cause a change in the heat-transfer characteristics, the effect depending on the tightness with which the sample is packed, whether channelling can occur, the volume of the gas and the rate at which it is produced. If the sample holder is too narrow, a close-packed sample can be pushed out of it by expanding gases released at the bottom of the holder. The previous thermal history of the sample may also be of importance, depending on whether any changes were reversible or not. Samples are sometimes examined "as received" and sometimes given a preliminary thermal treatment to make them comparable with other samples. Impurities which cause severe interference may need to be removed before investigation of the thermal behaviour of research samples.

The effect of packing is mainly on the rate of heat transfer. The more tightly the sample is packed the better the contact between individual particles and the more efficient the transfer of heat and the lower the temperature lag. The packing necessarily depends on the particle size of the sample. Another effect of the packing is on access of the atmosphere to individual particles. This depends on diffusion of gases into and out of the specimen, and so depends on the pore size between particles, and on the uniformity of packing. There will be a tortuosity factor on account if the many paths afforded the gas, and there will be diffusion of evolved gases throughout the specimen.

2.3. CRITERIA FOR THE EVALUATION OF A THERMOGRAVIMETRIC CURVE, AND DETERMINATION OF REACTION TEMPERATURE

Modern commercial thermobalances with variable heating rates, variable gaseous media, with vacuum or high-pressure facilities, and with continuous recording can satisfactorily deal with most of the problems discussed in the preceding section. Thermogravimetric curves should be evaluated on the basis of the chemical reaction under consideration. Among the commonest applications of the method in analytical chemistry, apart from the investigation of analytical precipitates (including behaviour on drying or ignition), is the so-called automatic gravimetric analysis introduced by Duval. Other applications are; following oxidation-reduction reactions of inorganic compounds, identifying new compounds in the study of solid state reactions, and studying thermal stability of organic compounds, mainly polymers.

In the study of analytical precipitates exposed to the effect of linearly increasing temperatures, the temperature intervals are determined in which a horizontal or slightly inclined part of the curve is obtained. However, as mentioned above, the position of these regions depends on several experimental conditions. With some compounds the horizontal part of the curve may be obtained at very low temperatures (often right from the beginning of the curve), e.g. $PbSO_4$ (from 20 °C), electrolytic silver (from 50 °C) and $PbCrO_4$ (from 110 °C). In other instances the horizontal part of the curve sets in from the beginning, but has a permanent slight incline due to slow oxidation, which may be irreversible, (for example in the case of electrolytically separated Cu and Cd), or reversible, when during cooling the curve returns to the starting point (precipitated Au). In some cases the curve may have a permanent slope, often appreciable, over its whole course, even in the regions where chemical reactions occur. Such a substance is not suitable for gravimetric determination because its composition in the whole temperature range is not constant (e.g.

$KLiFeO_6$ up to 947 °C, and the Ga cupferron complex up to 750 °C).
The most suitable case for TG evaluation is the curve which drops
continuously to its horizontal part corresponding to a stable compound.
This is the case with the majority of precipitated hydrated oxides when
they are converted into the anhydrous form. In this case it is as well to
make sure by another method (e.g. infrared spectroscopy) that the last
traces of moisture have disappeared. In the case of some salts and organic
complexes two horizontal sections are often formed on the TG curve. The
first horizontal zone corresponds to the weight of precipitate after
evaporation of the washing liquid, and the second corresponds to the
weight of the compound remaining after decomposition of the precipitate
and elimination of gaseous products. With hydrates containing a number
of bound water molecules it is possible to obtain, on slow heating, a TG
curve with a number of horizontal sections corresponding to single
dehydration steps (e.g. $NiSO_4 . 7 H_2O$). Very complicated curves displaying

losses and gains in weight are
observed in the case of a sul-
phide heated in the presence
of air, which results in the
formation of sulphates and
thus increases the weight. The
use of such compounds in
gravimetric analysis is unsatisfactory for these reasons.

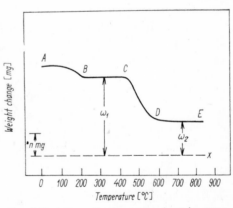

The method of automatic
gravimetric analysis developed
by Duval [27] is intended primarily for analysing precipitates of inorganic salts and complexes and it is based on an
accurate knowledge of the TG

Fig. 2.13 Automatic thermogravimetric curve
for a general, one-component system.

curve of the compounds concerned. It can be used in those cases where
the curve has a defined horizontal part. The principle of this method
can be explained by considering the example of the curve shown
in Fig. 2.13, representing a general TG curve with two horizontal
sections. The dotted baseline represents the level registered by the
thermobalance corresponding to a clean, dry crucible. It is evident that
after the evaporation of the wash-liquid the first plateau BC is formed
corresponding to weight w_1. After reaction, the second plateau DE is
formed corresponding to weight w_2. These plateaux indicate the formation
of stoichiometric compounds, and on multiplying the weight change by

the corresponding gravimetric analysis factor, it is possible to determine the amount of the corresponding metal ion. The method is rapid, requires a minimum of work, and shortens appreciably the time necessary for determination. However, it does require a knowledge of the experimental conditions under which the given precipitate forms stoichiometric compounds with well-defined horizontal sections on the TG curve. The method also permits the analysis of mixtures of two or more substances, for example a mixture of Ca and Mg oxalates, Cu and Ag nitrates, etc. The potential of automatic gravimetric analysis is considerable, as a result of the number of possible analytical precipitates mentioned in the literature. Duval gave about 500 examples of inorganic and organic compounds found suitable for automatic gravimetric analysis. In view of the limited size of this publication it is impossible to give examples of all possible applications of thermogravimetry in analytical, pure inorganic and organic chemistry. However, in addition to determination of stoichiometric compounds, the region of thermal stability and the so-called decomposition temperature can also be determined. The concept of the decomposition temperature, which is used as a characteristic property of most inorganic compounds, should, however, be employed with caution as its value depends on experimental conditions. The name "decomposition temperature" used in thermogravimetry is not a well chosen one, because it does not represent the true temperature of the decomposition reaction, but rather the lowest temperature at which the beginning of the change in the sample weight is observed with a particular apparatus. It has been shown earlier that it is experimental conditions, and not the absolute decomposition temperature, which define the temperature interval in which the reaction rate is zero. On a one-step TG curve the temperature when the weight begins to change and the temperature when the weight change is a maximum can be determined. The temperature corresponding to the peaks on the derivative TG curve does not correspond to the decomposition temperature either, and is not suitable for the determination of standard materials, as DTA is [82]. The lowest temperature at which a change in sample weight is recorded is called the "procedural decomposition temperature" [25]. From the literature it is evident that thermogravimetry is also useful in thermal treatment of polymers. However, in this case a number of irregularities are met, owing to the structural properties of polymers. The irregularities include slower reaction, and relatively small and well-defined temperature effects. Polymers do not display such sharp transition stages and the transitions are less dependent on the heating rate. Mechanical properties of polymers also appreciably affect the course of the TG curve. Therefore, characterization of the thermal behaviour of polymers and comparison based on thermal stability and decomposition temperature is not very good. Doyle [25] introduced the concept of the so-called "integral procedural decomposition temperature" and in so doing he put the widely differing behaviour of various polymers on a common basis, enabling comparisons to be made. This temperature is derived from the TG curve under strictly controlled experimental conditions. Doyle uses the "differential procedural decomposition temperature" (abbreviation: dpdt) for the determination of the inflexion point on the normalized TG curve, and the "integral procedural decomposition temperature" (abbreviation: ipdt).

The ipdt is determined from the TG curve as follows (Fig. 2.14). The area under the curve is expressed as a fraction of the total area of the plot and is thus normalized with respect to both the residual weight and the temperature. This area, A, is converted into an arbitrary temperature, T_A, by the following relationship

$$T_A = 875A + 25 \quad (2.5)$$

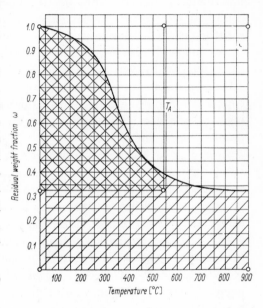

Fig. 2.14 Deduction of the value of the integral procedural temperature according to Doyle. Area of A: ////, area of K: XXXX. (Reprinted from [25] by permission of the copyright holder)

T_A represents an imaginary temperature at which substances which volatilize below 900 °C volatilize sharply. However, this is not a fair basis for comparing materials which may contain different amounts of refractory material. This deficiency may be corrected by a factor which takes into account the degree to which real materials approach the ideality assumed in T_A. This factor, K, is the ratio of the area under the curve from 25 °C to T_A °C to the area of the rectangle bounded by T_A and the residual weight fraction. The ipdt is given by the relationship

$$\text{ipdt} = 875K\,A + 25 \quad (2.6)$$

2.4. TEMPERATURE MEASUREMENT

It is evident from the preceding sections that one of the most important stages in the evaluation of a TG curve is the correct calibration of the temperature axis. The most frequently employed method for measurement of temperature changes in thermogravimetry is by means of thermocouples [47]. Although it is clear from the preceding sections that the site at which the temperature is measured should be actually in the sample or in its immediate proximity, this is not the case in some thermobalances, for constructional reasons. In this respect deflection balance systems are less suitable than null-point systems. The placing of the hot junction of the thermocouple directly in the sample is complicated by the possibility of contamination by volatile materials liberated in the reactions (e.g. S, HF, As). As a correct determination of temperature is an important condition for drawing correct conclusions from thermal analytical recordings, the principles of temperature measurement will be discussed in some detail.

According to Maxwell, the temperature of a body defines its thermal state considered from the point of view of its ability to transmit heat to other bodies. This is a quantitative expression and, therefore, it is necessary to consider temperature as a derived quantity, on the basis of energy and entropy. Direct measurement of the temperature as a physical quantity is accompanied by fundamental difficulties, because for real (non-ideal) bodies the temperature cannot be defined on the basis of physical properties. Therefore, calculated values are used which are a unique function of the temperature. Kelvin defined the absolute temperature scale on the basis of the expansion of an ideal gas. The Celsius temperature scale defines the temperature unit as a hundredth part of the temperature interval between the melting point of ice and the boiling point of water, corresponding originally to the temperatures 100 °C and 0 °C respectively (but now to 0 °C and 100 °C: it is not generally realized that Celsius "inverted" the scale). On Kelvin's absolute scale these two temperatures correspond to the values 273.15 K and 373.15 K. The temperature intervals 1 °C and 1 K are identical. Fahrenheit's temperature scale defines the melting point of ice as +32 °F and the boiling point of water as 212 °F. This scale is still employed in scientific papers in the U.S.A. For a more detailed study of the general expression of temperature, special literature is recommended [cf. 118].

A variety of thermometers exists for measuring temperature, e.g. solid expansion thermometers, liquid expansion thermometers, thermometers based on vapour pressure, electrical resistance thermometers, thermocouples, optical pyrometers, and radiation pyrometers. In thermal analysis the most commonly used method of temperature measurement is the thermocouple. Materials used for thermocouples should display as large a thermal e.m.f. as possible with increasing temperature, and these e.m.f.'s should be constant even on prolonged heating. The materials should also be very resistant to physical and chemical influences and they must be available in wire form. The thermoelectric potential depends on their degree of purity. Thermoelectrical inhomogeneity may be caused by the absorption of vapours of other metals (especially in the case of platinum), or by mechanical stress. With prolonged heating at elevated temperatures some thermocouple materials may be oxidized (e.g. $NiCr - Ni$), leading to permanently high results. Heating to glowing in a hydrogen atmosphere will regenerate the thermocouple in such cases. The thermocouple wires should be as thin as possible to minimize loss of heat. In the case of precious metals the diameters are usually $0.35 - 0.5$ mm, in other cases $0.5 - 3.0$ mm. To obtain a constant thermoelectric potential the wires should be artificially aged by electrical annealing. The method of making the junction depends on the materials used, but in general the following methods are used: soft-soldering up to 150 °C, hard-soldering up to 700 °C, and, at higher temperatures, welding in a reducing atmosphere. The welding is usually done with an electric arc or an oxygen - hydrogen flame. In arc welding a mercury or graphite arc is suitable and a flux is used, usually tricresyl phosphate. For thermocouples made from platinum metals, the most common in thermal analysis, graphite arc welding is suitable. In this case the mere contact of the end of the connected thermocouple wires (connected to one conducting wire of the current source) with the second electrode (normally a graphite rod) is usually sufficient. Welding by the discharge of condensers has been found suitable, permitting the direct formation of small welded sites between the two metals, one of which may be the crucible itself. Thermocouples are useful in thermal analysis because they are small and can be made in a form suitable for their intended positions. Care should be taken to avoid heat loss and that the response of the thermocouple is sufficiently rapid for the rate of temperature change used. The response depends on the size of the weld and the diameter of the thermocouple wires.

Every combination of metals or alloys gives rise to a thermoelectrical potential, but only certain combinations are of practical importance. The factors involved in the choice of a suitable combination are the magnitude of the thermoelectric potential, high electrical conductivity, low thermal coefficient of resistance, and good mechanical properties. The most commonly used combinations are tabulated in Table 2.1.

Table 2.1

Properties of Some Commonly Used Thermocouples

Type of thermocouple	Temperature range °C	E.m.f. at maximum temperature, mV	Remarks
Copper-Constantan (Cu 100%)-(Cu 55%, Ni 45%)	−200 to +600	32.04	oxidizing gases interfere
Silver-Constantan (Ag 100%)-(Cu 55%, Ni 45%)	−200 to +600	34.0	
Silver-Palladium (Ag 100%)-(Pd 100%)	+20 to +600	13.114	
Nichrome-Constantan (Ni 87.5%, Cr 12.5%)-(Cu 55%, Ni 45%)	+20 to +800	60.0	reducing atmospheres, SO_2, are harmful oxidizes above 700 °C
Iron-Constantan (Fe 100%)-(Cu 55%, Ni 45%)	+20 to +800	45.0	oxidizes above 600 °C
Nichrome-Nickel (Ni 87.5%, Cr 12.5%)-(Ni 100%)	+200 to +1200	45.0	reducing atmospheres, SO_2, are harmful, oxidizes above 800 °C
Nickel-Nickel iron (Ni 100%)-(Ni 66%, Fe 34%)	+200 to +1000	27.0	may be used without compensation leads
Platinum-Platinum rhenium (Pt 100%)-(Pt 92%, Re 8%) Le Chatelier's thermocouple	+20 to +1300	60.38	sensitive to S, C, Si, P
Platinum-Platinum rhodium (Pt 100%)-(Pt 90%, Rh 10%)	+20 to +1700	17.90	at elevated temperatures Rh sublimes
Gold palladium-Platinum rhodium (Au 60%, Pd 40%)+(Pt 90%, Rh 10%)	+20 to +1300	60.38	
Iridium-Ruthenium iridium (Ir 100%)-(Ru 10%, Ir 90%)	+20 to +1800	4.19	
Iridium-Iridium rhodium (Ir 100%)-(Ir 40%, Rh 60%)	+2000	11.85	
Tungsten-Tungsten molybdenum (W 100%)-(W 75%, Mo 25%)	+2700	6.0	

The relationship between thermoelectric potential and temperature for some common thermocouples is shown in Fig. 2.15 and the e.m.f.'s of some important thermocouples are given in Appendix No. 1.

The important properties of some thermocouples are given below. *Cu-constantan*; this is attacked by oxidizing gases and its e.m.f. depends on the purity of the copper used. It can be used from −200 °C upwards. *Ag-constantan*; this is very stable and it can be used in the range from −200 to +600 °C.

NiCr − Ni; this is relatively resistant to oxidizing gases. Owing to impurities Ni becomes brittle after prolonged use in the range 600 − 800 °C, and alloys containing 96% of Ni and small amounts of Al, Si, and Mn are used. Such thermocouples are known as chromel-alumel or Hoskin's thermocouples. When used con-

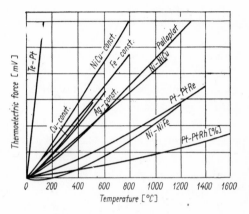

Fig. 2.15 Temperature dependences of most common thermocouples.

stantly at a temperature of over 800 °C an increase in e.m.f. $(1 − 2\%)$ is observed, owing to oxidation. The thermocouple may be regenerated by heating to glowing in hydrogen. At elevated temperatures it is unstable towards reducing and sulphurous gases.

NiFe − Ni; this is unstable in reducing, oxidizing and sulphurous atmospheres. At temperatures over 800 °C it should be protected against air oxidation. Between 600 and 800 °C it becomes brittle. Up to 100 °C its e.m.f. is approximately zero.

PtRe − Pt: this is very sensitive to reducing gases and gases containing S, CO, Si, and P. In an oxidative atmosphere it is stable up to 1200 °C. It displays a high e.m.f. value (approx. 60.0 mV at 1300 °C) as does the AuPd − PtRh thermocouple.

Pt10% Rh − Pt; this is called Le Chatelier's thermocouple and it is the most frequently used thermocouple. It is employed for measurements up to 1600 °C. It is sensitive to a reducing atmosphere in the presence of Si.

Pt30% Rh − Pt6% Rh; this is used for temperatures of up to 1800 °C where its e.m.f. is 13.25 mV.

For still higher temperatures, thermocouples of Ir − IrRu, W − WMo, and special combinations of graphite and carbides are used. These are not described here because so far they are of little importance in thermal analysis.

There are also special thermocouples which are not made by a simple welding of wires but by direct welding between the crucible and its carrier, e.g. the special sample carrier of the Mettler thermobalance (see Chap. 5).

The thermocouple wires should be insulated, especially in the region where they are exposed to high temperature. This insulation should be made with ceramic material. In the case of precious metal thermocouples, the insulating materials employed should not contain Si or S, because at high temperatures in a reducing atmosphere they form alloys with the thermocouple metal. The reliability of the measurement of the temperature with a thermocouple depends primarily on the method of protection against contamination. Therefore, the choice of the protecting ceramic material is of the utmost importance. For temperatures up to 900 °C, quartz capillaries may be used, even in a reducing atmosphere. The commonest insulators are ceramic materials of which the main component is mullite ($3 Al_2O_3 . 2 SiO_2$). However, in the case of this material and a reducing atmosphere at high temperatures, there is a danger of contamination from elemental silicon. Up to 1600 °C recrystallized materials containing predominantly Al_2O_3 are used; these are suitable in a reducing atmosphere, but have only an average resistance to thermal shock. Ceramic materials based on BeO, MgO, ThO_2 have approximately the same properties. The so-called stabilized ZrO_2 may be used in an oxidizing atmosphere up to 2400 °C in a reducing atmosphere and up to 2200 °C in vacuum. The platinum-rhodium thermocouple is very sensitive to contamination from silicon, iron, nickel, phosphorus, zinc, and sulphur, especially in a reducing atmosphere in which these elements are very reactive. In such cases hermetically sealed insulators, and thorough annealing at temperatures higher than the maximum temperature of measurement, must be used. A special case is the reduction of the ceramic material by platinum, especially in vacuum.

The value of the thermal e.m.f. at a given temperature of the hot junction depends on the temperature of the cold junction. As calibrated thermocouples are used, only when the temperature of the cold junction during measurement is identical with the temperature during calibration is the measured value correct. When the temperatures are different a correction may be made according to the relationship below:

$$E_s = E_m + \Delta E = E_m + k(T_m + T_k) \tag{2.7}$$

where E_s is the true e.m.f. value and E_m the measured value, k a constant for a given thermocouple, T_k the temperature of the cold junction during the calibration and T_m the temperature of the cold junction during the measurement.

The value of the constant k for Le Chatelier's thermocouple is 0.006 when the temperature of the cold junction is 20 °C. In the case of NiCr/Ni k is 0.041 and for NiCr/constant it is 0.062. Corrections for the temperature of the cold junction of Le Chatelier's thermocouple are given in Appendix No. 1. The cold junction may be maintained at a constant known temperature by immersing it in a small Dewar flask containing thawing ice (0 °C) or a thermostat with condensing water vapour (100 °C), or a thermostatically-controlled oil- or water-bath. The leads are made of the same material as the thermocouple arms, or, in the case of thermocouples made of precious metals, of different material showing identical thermoelectric potential.

Another method of correcting for the temperature of the cold junction consists of a compensation connection to a source of auxiliary current, in which the voltage is increased simultaneously with increasing temperature of the cold junction, which simultaneously produces an increasing thermoelectric e.m.f. The circuit is a Wheatstone bridge arrangement with a temperature sensitive resistance in one arm and in the opposite arm the thermocouple is connected in series with the measuring device (see Fig. 2.16). The description of other methods of compensation is beyond the scope of this book.

The actual measurement of the thermal e.m.f. is made with a galvanometer or millivoltmeter, or by a compensation method. The second type of measurement is more accurate and it is used in all instances where a very accurate temperature measurement is required. The simplest measuring device is a millivoltmeter without an auxiliary current source. In this case it should be borne in mind that the measured potential depends on the resistance of the thermocouple itself, including the leads to the measuring device and the internal resistance of the measuring device itself. The actual e.m.f. of the thermocouple is then given by the expression

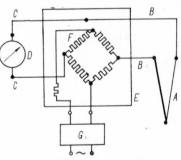

Fig. 2.16 Scheme with auxiliary voltage for the elimination of the effect of temperature of the thermocouple cold end.

A — thermocouple: B — compensation leads: C — copper leads: D — measuring device: E — connecting place at constant temperature with switch and temperature sensitive resistance F: G — source of auxiliary current

$$E = E_m \frac{R_t + R_v + R_g + R_c}{R_g} \tag{2.8}$$

where E_m is the voltage shown on the measuring device, R_t is the thermocouple resistance, R_v is the resistance of the leads to the measuring device, R_g is the internal resistance of the galvanometer (measuring device), R_c is the resistance of the compensating leads, and

$$E_m = E - i(R_t + R_v) \tag{2.9}$$

where i is the current through the galvanometer.

From this relationship it follows that a measuring device with a relatively high internal resistance should be used. The ratio of the internal resistance of the measuring device to the sum of the resistances of the leads should be 200 : 1. At such a ratio, changes in the resistance of the

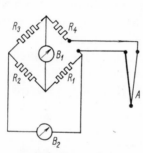

Fig. 2.17 Method of connection of two measuring devices to the thermocouple. A — thermocouple: B_1 and B — measuring devices: R_1, R_2, R_3, R_4: — resistances

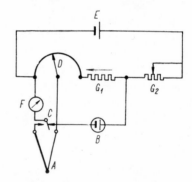

Fig. 2.18 Example of a compensation circuit for the measurement of the e.m.f. of the thermocouple.

A — thermocouple: B — standard electric cell: C — switch: D — potentiometer$_2$: E — dry cell: F — measuring devices: G_1 — resistance: G_2 — rheostat

leads, caused by corrosion of the thermocouples, are negligible. If the ratio is low a fixed resistance of the whole thermocouple branch should be taken into account, calculated from the known specific resistance of the materials used.

Other measuring devices such as a recorder are often used. In this case, it should be noted that an increase in current causes a decrease in voltage. This effect is smaller the smaller the thermocouple resistance and the higher the resistance of the measuring device. A second measuring device can be connected by the bridge arrangement shown in Fig. 2.17.

Single bridge resistances are used in order to keep the current in one measuring device independent of the current in the second.

The second method of measurement of the thermocouple e.m.f. is the compensation or null-point method. This method gives more precise results and eliminates the effect of the thermocouple resistance, as well as that of the leads. It consists of opposing a known, measurable, potential against the thermocouple e.m.f. until a zero-current state, indicated by the galvanometer, is achieved. One method of achieving this zero-current state is to use a potentiometer; a suitable circuit is shown in Fig. 2.18. Measurements are also made with recorders with built-in amplifiers, enabling small temperature changes to be measured even at high temperatures.

For temperature measurements in thermogravimetry, resistance thermometers may be used in addition to thermocouples. The principle of this method consists in the fact that the resistance of metals and semiconductors changes with temperature. In spite of the fact that this method is very precise, it is less suitable than a thermocouple because of the larger size of the thermal indicator, which is unsuitable for direct contact with the crucible or the sample. Resistance thermometers are more suitable for the measurement of the average temperature in a non-uniform temperature region. This method will not be discussed in detail but it suffices to note that the platinum resistance thermometer is the most commonly used, because it may be used up to a temperature of 1000 °C. The thermometer is composed of a resistance wire wound round a carrier and connected to the measuring device. Measurement of the resistance is made by the null-point method, using Wheatstone bridges of various types, or by the deflection method, in which the bridge is not balanced.

2.4.1. Special methods of temperature measurement in thermogravimetry

In addition to the methods of temperature measurement described in the preceding section, there are also other methods described in the literature, based on different principles. The need for different methods is due primarily to the fact that the use of thermocouples in direct contact with the sample or the crucible placed on the balance suspension affects the precision of the balance as the reference ends of the thermocouple must be connected to the measuring device. This connection is usually made of thin copper wire (0.05 mm) or wire of thermocouple materials, but in certain cases, such as the spring thermobalance or some microbalances, it is impossible without a negative effect on the precision of the balance.

Thus, other methods were devised, e.g. Marche and Carrol [64] built a transistor oscillator into the thermobalance suspension, containing a thermistor serving as a resistance probe for measuring the crucible temperature; Terrey [102] fixed into the thermobalance suspension (Gregg's balance) a miniature mirror galvanometer (see Fig. 2.19), which measured temperatures up to 900° with an error of ±10 °C, and Chatfield [48] measured the sample temperature directly by means of a thermocouple introduced through the balance suspension, avoiding the interfering effect caused by the connecting wires by doubling the beam edges and the beam support planes, and making them from a conducting material, thus forming the connection between the thermocouple arms and the measuring device. In fact the beam edges and supports were made of a hard chromium steel and polished to optical quality. A change in the thermocouple resistance in the interval $10-300\,\Omega$ has no substantial effect on the temperature recorded with a potentiometer; this is due to poor electrical conduction by the edges during oscillations of the beam. A very interesting method of temperature measurement in thermogravimetry was proposed by Šesták [96] who used the so-called two-crucible method; one

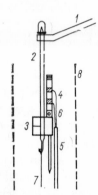

Fig. 2.19 Method of temperature measurement by a galvanometer suspended on the balance suspension. 1 — balance beam, 2 — glass suspension, 3 — magnetic system of the galvanometer, 4 — galvanometer, 5 — thermocouple, 6 — mirror of the galvanometer, 7 — balance suspension, 8 — balance jacket. (Reprinted from [102] by permission of the copyright holder)

with the sample (which is weighed) and the other with the thermometer. Both crucibles are blackened by electrolytically precipitated platinum black in order to ensure an equal emissive power of the surface. The thermometer crucible is slightly taller than the sample crucible, but its total surface area is made identical with that of the first crucible by cuts in the crucible wall. It is placed in the same temperature zone and a thermocouple is fixed to its base. In some cases the crucible itself may form part of the thermocouple. This arrangement permits accurate temperature measurement in vacuum and it is convenient in cases where the introduction of the thermocouple directly into the crucible with the sample or onto the sample wall is not possible. In certain cases, as for example in the Perkin-Elmer thermobalance [50], the temperature is not measured during the TG curve recording, but is evaluated on the basis of a known linear temperature programme.

In concluding this section, it should be stressed that the currently used method of temperature measurement, a thermocouple close to the sample, is a little dubious. If heating occurs by conduction or convection, the error in the measured temperature is usually less than 5%. The error depends on the position of the thermocouple and on other factors, such as the heat capacity of the thermocouple and the crucible. If heat transfer occurs by radiation, usual at high temperatures and in vacuum, the difference between the thermocouple and the crucible temperatures will depend mainly on their geometrical similarity, heat capacity, degree of surface blackening, quantity of heat removed by the thermocouples, etc. According to the Stefan – Boltzmann law the amount of irradiated or absorbed energy is proportional to the fourth power of the temperature, to the irradiated area, and to the degree of its blackening. Thus the conditions during single thermogravimetric measurements are not always quite reproducible. For most applications of thermal analysis the use of suitably chosen thermocouples is considered convenient, although for special cases, such as high-temperature vacuum thermogravimetry, where reproducibility of the temperature measurements is doubtful, another method of measurement should be chosen (e.g. a radiation pyrometer), at least as a control method.

2.5. TEMPERATURE AND BALANCE CALIBRATION

For practical temperature measurement an international temperature scale with further fixed points has been introduced. These points extend the temperature range beyond the limits set by the two fixed points of the Celsius scale. In the lower range of temperature the fixed points are determined by means of a resistance thermometer made of purest platinum. This thermometer was developed by Callendar [15] and it is characterized by the value 1.3910 for the R_{100}/R_0 ratio, where R_{100} is the electrical resistance at the boiling point of water at 760 mmHg, and R_0 the resistance at the melting point of ice. The central part of the temperature scale is defined on the basis of Le Chatelier's work [58] with the Pt/Pt10 % Rh thermocouple. The purity of platinum for this thermocouple should correspond to the temperature coefficient of resistance $R_{100}/R_0 = 1.3910$ (more recent papers require the value 1.3920). The thermocouple must fulfil the following limiting conditions:

1. $E_{Au} = 10300 \pm 50 \, \mu V$

2. $E_{Au} - E_{Ag} = 1185 + 0.158(E_{Au} - 10310) \pm 3\,\mu V.$

3. $E_{Au} - E_{Sb} = 4476 + 0.631(E_{Au} - 10310) \pm 5\,\mu V.$

where $E_{Au,Ag,Sb}$ is the potential of the thermocouple, with the cold end at 0 °C, at the solidification points of Au, Ag, Sb. The solidification point of Sb is a fixed secondary point on the international scale for the temperature 630.5 °C. The solidification of gold is assigned the temperature 1063 °C on the international scale and is measured by an optical pyrometer which has as its main advantage the absence of direct contact between the indicator and the heated body, which is a prerequisite of good reproducibility. In Tables 2.2 and 2.3 fixed points of the international temperature scale are given [56].

Table 2.2

Fixed Points of the International Temperature Scale

Primary fixed point	Temperature* °C	Reproducibility °C	Method of measurement
1. Boiling point of oxygen	−182.97	±0.002	Pt resistance thermometer
2. Melting point of ice	0.0	±0.0001	Pt resistance thermometer
3. Boiling point of water	100.00	±0.001	Pt resistance thermometer
4. Boiling point of sulphur	444.60	±0.005	Pt resistance thermometer
5. Solidification point of silver	960.80	±0.1	Pt-Pt (10%) Rh thermocouple
6. Solidification point of gold	1063.0	±0.1	Pt-Pt (10%) Rh thermocouple
7. Solidification point of gold	1063.0	±0.2	Optical pyrometer

* At 1 atmosphere pressure.

A thermocouple is calibrated by keeping its hot junction, at the moment of measurement, in a bath of solidifying metal, while the cold end is kept at 0 °C. The experimental conditions should enable the time of solidification to be several times longer than the time necessary for the measurement of the e.m.f. The measurement itself is made by the zero-current method. Another method of calibration is the so-called wire-method in which the disconnected hot end of a thermocouple is connected to a wire made from the calibration metal. This thermocouple, connected to a milli-voltmeter, indicates the moment of melting of the wire, while the thermo-

Table 2.3

Secondary Fixed Points of the International Temperature Scale

Secondary fixed point	Temperature* °C
1. Solidification point of Hg	—38.87
2. Triple point of water	0.0100
3. Decomposition temperature of $Na_2SO_4 . 10 H_2O$	32.38
4. Triple point of benzoic acid	122.36
5. Boiling point of naphthalene	218.0
6. Solidification point of Sn	231.9
7. Boiling point of benzophenone	305.9
8. Solidification point of Cd	320.9
9. Solidification point of Pb	327.3
10. Boiling point of Hg	356.58
11. Solidification point of Zn	419.5
12. Solidification point of Sb	630.5
13. Solidification point of Al	660.1
14. Solidification point of Cu	1083.0
15. Solidification point of Ni	1453.0
16. Solidification point of Co	1492.0
17. Solidification point of Pd	1552.0
18. Solidification point of Pt	1769.0
19. Solidification point of Rh	1960.0
20. Solidification point of Ir	2443.0
21. Solidification point of Mo	3380.0

* At 1 atmosphere pressure.

couple to be calibrated is placed in its immediate proximity. In the case of the metallic bath method, materials listed in Table 2.4 are used, for which preparative methods have been devised guaranteeing a high degree of purity.

In view of the fact that these calibration methods take a good deal of time and work, thermocouples calibrated by the reference method are used in current practice. This method consists in the comparison of the thermal e.m.f. with that of a thermocouple calibrated by some of the methods mentioned, which thus serves as a standard. It is recommended that in thermogravimetric measurements in which thermocouples are used, these be calibrated in advance. In view of the danger of contamination and changes in the general experimental arrangement the calibration should be repeated from time to time. According to Duval [27] the calibration should be performed once a year. However, as these methods of calibration are rather lengthy, and because the rapid spread of thermogravimetric methods has led to the introduction of various types of apparatus with different methods of temperature measurement, it was found necessary to introduce other methods of checking the accuracy of the temperature measured, especially since comparison of the so-called decomposition

Table 2.4

Temperature Calibration by the Bath Method

Contents of bath	Temperature °C	Vessel	Atmosphere	Covering layer	Remarks
1. Ice + water	0.0	Dewar	air	paraffin	saturated with water vapour
2. Tin	231.9	porcelain, graphite	reducing	graphite	easily undercooled
3. Cadmium	320.9	quartz, graphite, alumina	reducing	graphite	poisonous vapour
4. Lead	327.3	porcelain, graphite	reducing	graphite	
5. Zinc	419.5	quartz, graphite	reducing	graphite	
6. Antimony	630.5	graphite	reducing	graphite	
7. Aluminium	660.1	graphite	argon	graphite	work in Si ceramics impossible
8. Silver	960.5	graphite	reducing	charcoal	melting point fluctuates $\pm 0.1\,°C$
9. Gold	1063.0	alumina	air		
10. Copper	1083.0	graphite	air	charcoal	melting point fluctuates $\pm 0.1\,°C$
11. Nickel	1453.0	alumina	argon, vacuum	glass, $BaCl_2$	
12. Cobalt	1492.0	alumina	argon, vacuum		
13. Palladium	1552.0	alumina	argon, vacuum		

temperatures and other data derived from the TG curve showed a lack of agreement between the results obtained with different apparatus. One method is the use of "temperature standards", i.e. substances having suitable thermal characteristics. Duval [27] uses, for example, calcium oxalate as a temperature standard for calibration after each new adjustment of the thermobalance. The use of substances with a well-defined course of the thermogravimetric curve was investigated by a number of workers [33, 85, 110], who also examined the possibility of using the peaks of the derivative TG curve and DTA curve. This question is being studied by the standardization committee of ICTA. As mentioned earlier, the concept of "decomposition temperature" used in connection with the thermogravimetric curve is a complex one depending on several experimental conditions. The concept of the so-called "procedural decomposition temperature" [25] was introduced, which on the basis of later work [53], was considered the most suitable parameter in investigations of substances suitable as temperature standards in thermogravimetry. Approximately 30 substances were selected which can be used as thermogravimetric temperature standards under strictly observed experimental conditions. Their properties are tabulated in Chapter 4. Thus the balance itself can be calibrated for its sensitivity and precision, as well as its weighing limits, simultaneously with the thermometer system.

An original method of temperature calibration in thermogravimetry was introduced by the Perkin–Elmer company [50]. This method is based on the use of reversible magnetic transitions in ferromagnetic materials.

Table 2.5

**Substances Used for Temperature Calibration
from their Reversible Magnetic Transitions**

Material	Curie temperature °C
Alumel	160
Nickel	360
Iron	780
"Hisat 50" alloy	960

This new method of calibration was made possible by the construction of a miniature electric furnace consisting of a small heating ring surrounding the crucible only; this permitted a permanent magnet to be placed outside the furnace, with the weighed sample in its magnetic field. The calibration is made with small amounts of samples of various ferromagnetic substances.

These substances affect the balance beam by their magnetic force, and this force is recorded as a function of temperature. Every substance loses its ferromagnetic properties at a defined temperature (the Curie temperature) and this is manifested on the weight-change curve as a weight loss. Substan-

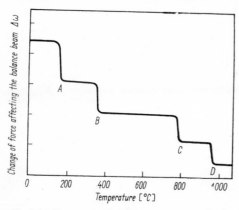

ces used for this type of temperature calibration are listed in Table 2.5. The main advantage of this method is that the calibration corresponds to the temperature of the position of the crucible and that in a single experiment several points on the temperature scale can be fixed. Also, the calibration is independent of the nature of the atmosphere and pressure, and the transitions are reversible. In Fig. 2.20 a typical record of the temperature calibration on the basis of magnetic changes is shown.

Fig. 2.20 Temperature calibration on the basis of the Curie point. A — alumel 160 °C, B — Nickel 360 °C, C — Iron 780 °C, D — Hisat 50 960 °C. (Reprinted from [50] by permission of the copyright holder)

2.6. SPECIAL METHODS OF THERMOGRAVIMETRY

In this section methods will be discussed which by various arrangements or by special construction of the apparatus produce both the thermogravimetric curve, and some other feature. The most interesting of these methods, which is now being increasingly used, is derivative thermogravimetry (DTG). In contrast to thermogravimetry which follows the change of the sample weight as a function of time or temperature, DTG follows the rate of weight change of the sample with respect to time or temperature $dw/dt = f(T \text{ or } t)$.

Both TG and DTG curves have been mentioned earlier and their shape was discussed with reference to Fig. 2.1. The shape of the derivative curve is similar to that of differential thermal analysis (DTA) which will be discussed in detail in the appropriate chapter. In contrast to the DTA curve it has the advantage that its baseline returns to zero after the reaction, which is often not the case with the DTA curve. However, its use is limited to reactions in which a sufficiently rapid change in weight takes place. The DTG curve may be obtained by differentiating the classical TG curve numerically, or automatically by using a differential thermobalance. Quantitative use of this method by the evaluation of the area under the curve is questionable because the area depends on a number of factors, such as the temperature-dependence of the reaction rate. Therefore, these curves are used mainly for qualitative purposes where they are very useful in identification of reactions within a narrow temperature range, superimposed on the normal TG curve. This method is a very useful supplement to both classical thermogravimetry and differential thermal analysis. Comparison permits the identification of reactions which are accompanied by a weight change. The use of this method is also possible in the analysis of melts. This method was applied for the first time in 1953 by de Keyser [55]. He heated two identical samples hung one on either

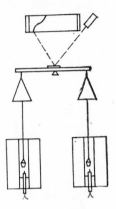

Fig. 2.21 Derivative thermobalance according to de Keyser. Schematic representation. (Reprinted from [55] by permission of the copyright holder)

end of an equal-arm beam, using a double furnace, one half of which was kept a few degrees (approx. 5 °C) hotter than the other during the temperature programme. The principle of this method is shown in Fig. 2.21. Equipment based on this principle is presently manufactured by the firm Sartorius (see Chap. 5). From the experimental point of view, however, this method is rather exacting, and it was developed and simplified by a number of workers [16, 75, 106]. The wide use of this method is mainly due to Erdey and the Paulik brothers [75] who obtained the derivative curve in a simple manner from the rate of vertical movement of the balance suspension. The apparatus developed by them, known as a "Derivatograph", is widely used. It records the DTA curve in addition to the TG and DTG curves. Their constructional device for obtaining the derivative TG curve is applicable only in the case of deflection-type balances (see Chap. 5). Some currently produced thermobalances give simultaneous recording of the DTG and TG curves.

Another method is fractional thermogravimetry. This method was introduced by Waters [107] in 1960. It is useful mainly in cases where several reactions are followed simultaneously. Its principle is that certain volatile compounds liberated during the thermal decomposition of the sample are selectively condensed or absorbed, and weighed together with the sample.

Non-absorbed components correspond to the recorded weight loss, and can either be moisture or gaseous components. The original design is shown schematically in Fig. 2.22. The sample carrier A is composed of a glass vessel connected via a capillary to the absorption flask C. The whole system, which is made of glass, is hung on the suspension of the thermobalance. Waters applied this method mainly to the analysis of coal, and was able to separate various pyrolytic fractions.

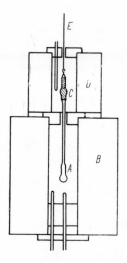

A similar method, known as inverse thermogravimetry, consists in following the weight increases of the absorbent on which the gaseous products formed by thermal decomposition of the sample are absorbed. The weight change of the sample is not recorded. In this case the sample is not placed on the suspension of the thermobalance but in a furnace separate from the balance system. The method permits the sample to be heated under accurately defined thermal conditions, gives easier and more accurate measurement of the sample temperature, and does not limit the sample weight. The absorbent will absorb the gaseous product rapidly. The development of this method is due mainly to Škramovský [101].

Fig. 2.22 Arrangement for fractional thermogravimetric analysis. A — bulb containing the sample, B — furnace, C — small absorption tube, D — small subsidiary furnace, E — balance suspension. (Reprinted from [107] by permission of the copyright holder)

There are also methods of thermal analysis under well-defined conditions of atmospheric composition and pressure (see pp. 28 – 32 for examples). In these cases experimental conditions are created under which the reactions investigated are reduction, oxidation, etc. It should be noted that the choice of experimental conditions, among which the effect of the atmosphere in the reaction space of the thermobalance should be included, may strongly influence the process under investigation. The aerodynamic or pressure conditions can influence the kinetics of the process. From this point of view, thermobalances working in a vacuum become of prime importance. Their construction will be dealt with in Chapter 5. The importance of vacuum thermobalances, especially microbalances, is shown by the fact that the National Bureau of Standards,

Washington, organized, in 1961, an independent conference devoted to this topic [115].

2.7. STUDY OF REACTION KINETICS BY THERMOGRAVIMETRY

Thermogravimetry, differential thermal analysis and other thermoanalytical methods (e.g. dilatometry and calorimetry) can be used to study the kinetics of a chemical reaction and determine basic kinetic constants such as the rate constant, activation energy, order of the reaction, and frequency factor. These methods usually measure continuously and automatically a change in some physical property, such as weight, enthalpy, length or the volume of the given system as a function of temperature. Only those methods which determine a change in weight will be discussed. These methods may be used for reactions characterized as follows [20, 62].

1. Decomposition reaction of the type

$$A_{(solid\ phase)} \rightarrow B_{(solid\ phase)} + C_{(gas\ phase)}$$

2. Reaction between two solid substances

$$A_{(solid\ phase)} + B_{(solid\ phase)} \rightarrow C_{(solid\ phase)} + D_{(gas\ phase)}$$

3. Reaction between a solid and a gas

$$A_{(solid\ phase)} + B_{(gas\ phase)} \rightarrow C_{(solid\ phase)}$$

4. The transition of a solid or a liquid substance to a gas

$$A_{(solid\ phase)}\ \text{or}\ A_{(liquid\ phase)} \rightarrow B_{(gas\ phase)}$$

When the kinetic study is based on observation of the weight change, two approaches are possible in principle, viz. the isothermal and the dynamic heating methods. The isothermal or static method is still the more commonly used. It is the determination of the degree of transformation at constant temperature as a function of time. Over the last ten years the use of the dynamic method has been increasing. The dynamic method is the determination of the degree of transformation as a function of time, during a linear increase of temperature. Methods and mathematical treatment of results are given in the papers by Šatava, Šesták, and Škvára [92, 93, 95, 97−99], Háber and co-workers [43], and others [24, 34, 36]. The static method is used more commonly and is considered by certain authors to be more accurate. Comparisons of the scope and the precision of both methods [93, 97, 98] showed comparable results with respect to precision. Generally, it can be said that the static method is more suitable for obtaining information about the slowest process, and also the reaction order and the reaction mechanism. The dynamic method is more suitable for obtaining data on the kinetics of the reaction from a single curve for the whole temperature range.

Static method

The conventional method of study of the kinetics of thermal decompositions of solids is based on observation of the reaction at constant temperature, the determination of the rate equation for the reaction course, and the determination of the dependence of the rate constant on temperature, from measurements at several temperatures under static conditions. Hence, the amount of product, α, is measured as a function of time t, at constant temperature T and pressure P. The reaction mechanism is then determined by testing the fit of the experimental results with the possible kinetic equations. Using the Arrhenius equation it is possible to calculate the values of the activation energy and of the frequency factor at various temperatures. In the experiment the sample is introduced into a preheated thermobalance, or a preheated furnace is pushed over the balance suspension with the sample. An example of a decomposition reaction is shown in Fig. 2.23.

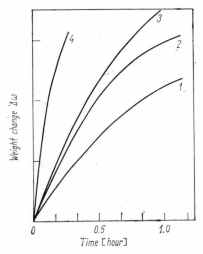

Fig. 2.23 Isothermal weight loss curves of a decomposition reaction, obtained at various temperatures, increasing from curve *1* to curve *4*. (Reprinted from [5] by permission of the copyright holder)

The mechanism of thermal decomposition reactions is given by the nucleation theory [62]. According to this theory, nuclei of substance $B_{(solid)}$ are formed at various sites in the structure of the original substance $A_{(solid)}$. The new phase B is formed from these nuclei. The time required for the formation of these nuclei in the reacting substance is called the induction period. The nuclei may be formed by heating. The main part of the isothermal weight change curve is a gradual loss in weight as a function of time.

In the study of kinetics in heterogeneous systems in the presence of a solid phase a dimensionless quantity known as the degree of "transformation" (in this case decomposition) is used, and given the symbol α, and is the ratio of the volume of the solid phase reacted at time t, to the original volume of the starting phase at time $t = 0$. The evaluation of kinetic data is difficult if the reaction mechanism is unknown. Some methods of evaluation are based on the supposition that the kinetics can be expressed by a formal kinetic equation [99]. Using this definition of the degree of transformation, this equation may be written in the following form:

$$d\alpha/dt = k(1 - \alpha)^n \qquad (2.10)$$

where t is time, k is the rate constant, and n is the order of the reaction. The kinetic equation may be more complex, however, and Zsako [117] deduced a more general form of this formal kinetic equation:

$$d\alpha/dt = k \cdot \alpha^a (1 - \alpha)^b \qquad (2.11)$$

where a and b are empirical constants.

From Eq. (2.10) it is evident that the mechanism of the reaction is reflected in the value of the reaction order n and the rate constant k. The dependence of the rate constant on temperature is given by the Arrhenius equation

$$k = Z \exp(-E/RT) \qquad (2.12)$$

where Z is the frequency factor and E the activation energy. From a plot of log k versus $1/T$, E may be determined (see Fig. 2.24) [5]. Maintaining a constant temperature requires a furnace of large mass, which usually takes a long time to heat. However, the greatest experimental difficulty is the introduction of the sample into the preheated furnace. Time elapses before the sample attains the required temperature and during this time the reaction

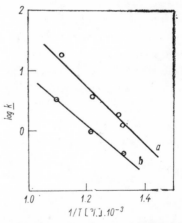

Fig. 2.24 Plot of log k versus $1/T$ (according to Arrhenius relation) for the reaction of uranium carbide and cyanide with carbon dioxide.
a: carbide — b: cyanide
(Reprinted from [5] by permission of the copyright holder)

occurs to a certain extent. Thus, instead of a correct value of the weight loss corresponding to the termination of the reaction, a value is obtained which is decreased by an amount corresponding to the reaction which took place before the required temperature was attained. This may have a great effect on the values of the kinetic constants. However, in the case of first-order decomposition where the rate constant does not depend on the initial sample weight, the error does not occur. In all other cases, i.e. when $n \neq 1$, the true value of the rate constant k is different from the measured k'. A sample holder of a low heat capacity does not help, as this causes self-heating or cooling as a result of enthalpy changes accompanying the decomposition reactions. One solution is the use of a thin foil "crucible" [93]. Such a "crucible", after introduction into the preheated furnace, is brought into contact with a conducting block of large heat capacity, pre-heated to the required temperature. This block serves as the holder of the "crucible". The disadvantage of a non-accurate determination of the beginning of the de composition reaction in the static method may be eliminated by the introduction of a correction at the beginning of the isothermal weight change curve [70, 103].

From the proposed elementary steps constituting the reaction, kinetic equations are deduced which are then tested with the results obtained. The equation which best describes the relationship between degree of trans-formation and time describes the slowest, i.e. rate-limiting, step. Kinetic equations are set up corresponding to one of three possible steps as the rate-limiting step. These rate-limiting steps are diffusion, nucleation, and reaction at the phase boundary.

One-dimensional diffusion with a constant diffusion coefficient follows the parabolic law according to the equation

$$\alpha^2 = (k/x^2) . t \tag{2.13}$$

where x is the distance of the boundary from the zero position.

Two-dimensional diffusion with cylindrical symmetry in a cylinder of radius r is given by Eq. (2.14) [111]:

$$(1 - \alpha) \ln(1 - \alpha) + \alpha = (k/r^2) . t \tag{2.14}$$

A more thorough analysis of this mechanism was made by Valensi [112−114]. When the volumes of the starting and product material are equal, the relationship is reduced to Eq. (2.14).

For three-dimensional diffusion of spherical symmetry Jander's equation [51] applies:

$$[1 - (1 - \alpha)^{1/3}]^2 = (k/r^2) . t \tag{2.15}$$

For a reaction starting from the surface of a spherical particle of radius r the equation of Ginstling and Brounshtein [41] expresses the process:

$$(1 - 2\alpha/3) - (1 - \alpha)^{2/3} = (k/r^2) . t \tag{2.16}$$

A more detailed analysis of this mechanism was carried out by Carter [17], Valensi [114] and others [116] who also took into consideration differences in volumes between the starting material and the product. If the volumes are equal, Valensi's relationship reduces to Eq. (2.16).

Certain reactions in the solid phase follow the first-order kinetic equation where $n = 1$. This includes the nucleation mechanism in which the rate of appearance of product is limited by the rate of growth of the nuclei. This model supposes a random nucleation of the product on active centres of the crystal structure, with only one nucleus formed on each particle [1, 2, 3]. The integrated first-order rate equation is:

$$\ln(1 - \alpha) = -kt \tag{2.17}$$

The growth of nuclei and the nucleation were studied by Avrami [2] and Erofèev [29] who deduced the following relationships for random nucleation:

$$\sqrt[2]{-\ln(1 - \alpha)} = kt \tag{2.18}$$

$$\sqrt[3]{-\ln(1 - \alpha)} = kt \tag{2.19}$$

The rate of appearance of the product may be controlled by reactions at the phase boundary. This is the case when diffusion through the layer of product is rapid compared with the reaction at the phase boundary. The product layer is not compact when its molar volume is substantially smaller than that of the starting substance on which it is formed. In this case the limiting step is the chemical reaction at the boundary. A process which depends on the size of the initial surface area is called topochemical. The equations deduced for this model on the basis of simple geometrical forms are based on the assumptions that the reaction rate is controlled by the phase boundary, that the reaction rate is proportional to the surface area of the unreacted part of the material, and that nucleation is instantaneous, as a result of which the surface of each particle is covered by a layer of the product. If the reaction is controlled by the movement of the boundary at constant rate u, then according to Sharp, Bridley and co-workers [11, 88], for spherical symmetry the following equation applies:

$$[1 - (1 - \alpha)^{1/3}] = (u/r) \cdot t \tag{2.20}$$

For cylindrical symmetry the equation is

$$[1 - (1 - \alpha)^{1/2}] = (u/r) \cdot t \tag{2.21}$$

where r is the radius of the sphere or the cylinder.

For certain values of the reaction order n (2/3, 1/2, 1), the integrated form of Eq. (2.10), expressed in the form

$$1/n - 1[1/(1 - \alpha)^{n-1} - 1] = kt \tag{2.22}$$

gives relationships mentioned above. Thus, with $n = 1/2$, Eq. (2.21) is obtained, and with $n = 2/3$, Eq. (2.20). Values of n different from $1/2$, $2/3$ and 1 cannot be accounted for by any of these three mechanisms as the rate-limiting step.

Dynamic method

As early as 1932 [89, 100] Škramovský indicated the possibility of using a thermogravimetric curve for the study of kinetic data. He also demonstrated its advantages compared with the static method then used. As the method demands strict experimental conditions and a high accuracy of results, development took place only after 1960 when the equipment reached a sufficiently high standard. Until then the thermogravimetric method was employed mostly for qualitative and quantitative analytical evaluation of chemical compounds.

Compared with the static method, the advantages are that it requires a far smaller number of experimental data, and kinetic values may be determined from a single thermogravimetric curve for the whole temperature range, and for the whole kinetic analysis one sample suffices. On the other hand, it should be remembered that the measured values (temperature, change of weight of the sample) should be obtained with the maximum possible accuracy, bearing in mind any effects of the experimental conditions. According to Šatava [94], the theoretical study of the application of a thermogravimetric curve for mathematical evaluation of kinetic data originates from Krevel and co-workers, and Freeman and Carroll [34]. The last-mentioned authors published the method for the treatment of TG curves of the decomposition of solids (such as calcium oxalate) without considering the effect of experimental conditions on the accuracy of the results obtained. A series of later papers was devoted to this problem [28, 37, 93, 94, 96−99].

Generally, the method consists of expressing the relationship between the "degree of change" (α) and time. The treatment can be carried out in two ways. In the first, various mechanisms with different rate-limiting steps are proposed and the results obtained tested with the resulting equation for each mechanism. The rate-limiting step (the slowest) is expressed by that equation which is in best agreement with the experimental values. However, as the mathematical treatment is complex this method is not commonly used for the evaluation of the TG curves [96], and a second, simpler method is used, which makes use of the general formal kinetic equation (2.10), mentioned earlier, in which the degree of transformation (α) is now defined as the ratio w/w_∞ (where w is the loss in weight at time t, and w_∞ is the maximum loss in weight on the TG curve). It should be borne in mind that the kinetic constants computed from this equation have only an empirical character. The mechanism of the reaction is deduced from the calculated reaction order and rate constant.

To solve the differential equation (2.10) three methods may be used, viz. differential, integral, and approximate. A full discussion of all these methods is beyond the scope of this book and therefore only some of the most often used will be presented here. A more thorough study is given in the literature [98, 99].

The most commonly used are differential methods, the work of Freeman and Carroll [34] being a typical example. This method is analogous to the differential method used in static measurement, in which the value of the rate constant k is obtained from the mean value of several values calculated from the isothermal curve. In this case a single value of the rate constant is obtained for each point on the TG curve. If the expression for k given by Eq. (2.12) (the Arrhenius equation) is substituted into Eq. (2.10), then

$$\frac{d\alpha}{dt} = Z \cdot e^{-E/RT}(1 - \alpha)^n \tag{2.23}$$

If the rate of temperature change is q, i.e.

$$\frac{dT}{dt} = q \tag{2.24}$$

On substituting for dt in Eq. (2.23), the following equation is obtained:

$$\frac{d\alpha}{dT} = \frac{Z}{q} \cdot e^{-E/RT}(1 - \alpha)^n \tag{2.25}$$

which on taking logarithms becomes

$$\ln\left(\frac{d\alpha}{dT}\right) = \ln\frac{Z}{q} - \frac{E}{RT} + n \cdot \ln(1 - \alpha) \tag{2.26}$$

and on differentiation the following expression is obtained:

$$d\ln\left(\frac{d\alpha}{dT}\right) = -\frac{E}{R} \cdot d\frac{1}{T} + n[d\ln(1 - \alpha)] \tag{2.27}$$

On dividing the whole equation by $d[\ln(1 - \alpha)$ the final relationship for the determination of kinetic values is obtained:

$$\frac{d\ln\left(\dfrac{d\alpha}{dT}\right)}{d\ln(1 - \alpha)} = -\frac{E}{R}\left[\frac{d\dfrac{1}{T}}{d\ln(1 - \alpha)}\right] + n \tag{2.28}$$

which can be written in a shortened form as

$$y = -\frac{E}{R} \cdot x + n \tag{2.29}$$

Equation (2.29) is usually represented graphically by plotting y against x, and the value of the activation energy E is determined from the slope of the straight line obtained. The value of the reaction order n is determined from the intercept on the x axis. It should be pointed out that Eq. (2.10) is only formal and the reaction order determined has no physical meaning.

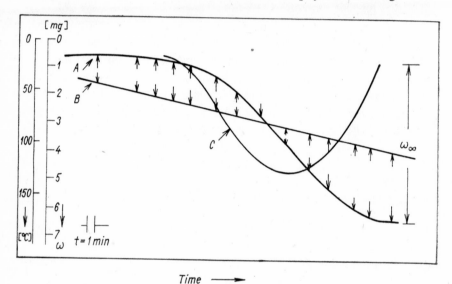

Fig. 2.25 Example of a TG curve and its evaluation by numerical differentiation. Sample: $\alpha - CaSO_4 \cdot 0.5 H_2O$.

Curve A: Plot of weight loss vs. time. Curve B: Plot of temperature vs. time. Curve C: Plot of derivative curve. Heating rate 1.7 °C/min. (Reprinted from [93] by permission of the copyright holder).

In practice the treatment of a TG curve such as that in Fig. 2.25 is to divide it into temperature intervals of 2 °C, read the corresponding weight changes, and determine the decomposition rates dw/dt. The actual reaction rate at a certain moment may be determined graphically, or, more accurately, by numerical differentiation. The dw/dt values are then plotted on the original TG curve and average values are read from the constructed curve. Evaluations based on various methods of numerical differentiation may be dealt with by computer [91, 96].

Other methods of determining kinetic data from the TG curve are the integration and approximation methods [99], of which a number are known. The method by Šatava and Škvára [95] should be noted.

Starting with the general kinetic equation for the decomposition of a solid:

$$\frac{d\alpha}{dt} = kf(\alpha), \tag{2.30}$$

where $f(\alpha)$ depends on the reaction mechanism, and combining this equation with Eqs. (2.12) and (2.24), the following relationship is obtained:

$$\frac{d\alpha}{f(\alpha)} = \frac{Z}{q} \cdot e^{-E/RT} \cdot dT \tag{2.31}$$

the integration of which gives the TG curve [92, 99, 117]. The left-hand side of the equation may be integrated if the function $f(\alpha)$ is known, which leads to a new function $g(\alpha)$. The right-hand side may be integrated with respect to T, if the value of the activation energy E is known. On integration of Eq. (2.31) the following equation is obtained:

$$g(\alpha) = \frac{ZE}{Rq} \cdot p(x) \tag{2.32}$$

where $p(x)$ is defined by the equation

$$p(x) = \frac{e^{-x}}{x} - \int_{x}^{\infty} \frac{e^{-u}}{u} \cdot du \tag{2.33}$$

where $u = E/RT$ and $x = u$ at the temperature of the sample. This function has been tabulated by some authors [25a, 117]. The logarithmic form of Eq. (2.32) is

$$\log \frac{ZE}{Rq} = \log g(\alpha) - \log p(x) = B \tag{2.34}$$

It is seen that B is independent of temperature. Thus, instead of Eqs. (2.10) or (2.11) which describe the decomposition only algebraically, the function $g(\alpha)$ corresponding to kinetic equations for possible reaction mechanisms is used. The kinetic analysis is also simplified by carrying out graphical comparison of the functions $\log g(\alpha)$ and $\log p(x)$ and plotting for various values of activation energies the expression $-\log p(x)$ versus T, for the temperature range $0-1000$ °C. The values of $\log g(\alpha)$ are computed for the most probable kinetic equations for the mechanisms of thermal decomposition of solid substances (diffusion, nucleation, reaction at the phase boundary) and tabulated in a manner similar to that of Sharp and co-workers [88]. The TG curve is analysed by finding T_α values for values of α at intervals of 0.05. The values of $\log g(\alpha)$ (tabulated for various reaction mechanisms as a function of corresponding T_α values) are then put on transparent paper on the same scale as the plot of $-\log p(x)$. The transparent paper with the $\log g(\alpha)$ plot is superimposed on the graph of $\log p(x)$; if the temperature scale coincides, then agreement with some of the $\log p(x)$ curves is sought. From this curve the value of the activation energy E is then determined. The agreement with some of the model curves of $\log g(\alpha)$ determines the most probable kinetic equation for the thermal decomposition. The procedure is shown in Fig. 2.26.

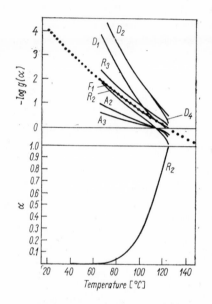

Fig. 2.26 Method of evaluation of the TG curve. The lower curve corresponds to function R_2, for $E = 20$ kcal/mole, $Z = 10^9$. s^{-1} . mole$^{1/2}$ at a heating rate of 1 °C/min. The upper curves are plots of log $g(\alpha)$ vs. T, calculated from the TG curve for various kinetic equations. The dotted curve is a plot of log p vs. T for $E = 20$ kcal/mole. Curves D_1 for Eq. (2.13), D_2 for Eq. (2.14), D_3 for Eq. (2.15), D_4 for Eq. (2.16), F_1 for Eq. (2.17), A_2 for Eq. (2.18), A_3 for Eq. (2.19), R_2 for Eq. (2.21), and R_3 for Eq. (2.20), (Reprinted from [95] by permission of the copyright holder)

In recent years TG curves have been increasingly used for the evaluation of kinetic data, not only in the field of polymer stability control [19, 31, 32, 79, 87], but also in the investigation of other reactions, such as dehydrations, redox reactions, decompositions, etc. [26, 46]. The deviations within published kinetic constants measured and calculated in different ways led to a careful study of experimental conditions and possible errors arising during these determinations, especially with the thermogravimetric method. The sources of error which may affect the results of this method were analysed by Šesták [98] who classified them into three categories.

(a) Errors in direct measurements, i.e. changes of weight and temperature. This problem is connected with the accuracy of the measuring devices, the controlling of factors affecting the TG curve, the method of temperature measurement, and the nature of the temperature programme.

(b) Precision with which the reaction conditions are maintained and the disturbing effects eliminated, including the effects of heat transfer, temperature gradient in the sample, enthalpy changes in the sample, and problems arising from mass transfer by internal or external diffusion.

(c) Precision in the mathematical treatment of the curves and results obtained from a single point on a curve. The errors for single point methods usually do not exceed $\pm 10\%$ for the activation energy value.

In conclusion it should be stressed that in all kinetic studies, both static or dynamic, the reason and the aim of the measurement should be specified. The correct choice of experimental conditions should ensure that the measurements are made on the required rate-limiting step. In comparing the two possible approaches, the static method undoubtedly gives more information on the reaction mechanism. The dynamic method may have certain advantages in the case of rapid reactions, and also gives a better indication of the state of the sample at any moment, with respect to temperature and degree of reaction. However, it is unlikely that the dynamic technique will ever find universal application, because the rate expression

$$\frac{d\alpha}{dT} = k(1 - \alpha)^n \tag{2.10}$$

is not applicable to all solid-state decompositions.

References

1. AVRAMI M. *Chem. Phys.* **7,** 1103 (1939).
2. AVRAMI M. *Chem. Phys.* **8,** 212 (1940).
3. AVRAMI M. *Chem. Phys.* **9,** 177 (1941).
4. BARRET P., PERRET R. *Bull. Soc. Chim. France* 1459 (1957).
5. BEŇADIK A., BLAŽEK A., EDEROVÁ J. *Collection Czech. Chem. Commun.* **35,** 1154 (1970).
6. BERGSTEIN A., VINTERA J. *Collection Czech. Chem. Commun.* **22,** 884 (1957).
7. BLAŽEK A., CÍSAŘ V., ČÁSLAVSKÁ V., ČÁSLAVSKÝ J. *Silikáty* **6,** 25 (1962).
8. BLAŽEK A. *Bergakademie* **12,** 191 (1960).
9. BLAŽEK A., CÍSAŘ V., ČÁSLAVSKÁ V., ČÁSLAVSKÝ J. *Collection Czech. Chem. Commun.* **25,** 2419 (1960).
10. BLAŽEK A., CÍSAŘ V. *Silikáty* **3,** 26 (1959).
11. BRINDLEY G. W., SHARP J. H., PATTERSON H. I., ACHER B. N. N. *Am. Mineralogist* **52,** 201 (1967).
12. BROWN D. H., NUTTALL R. H., SHARP D. W. A. *J. Inorg. Nucl. Chem.* **25,** 1067 (1963).
13. CAHN L., SCHULTZ H. *Anal. Chem.* **35,** 1729 (1963).
14. CAHN L., PETERSON N. C. *Anal. Chem.* **39,** 403 (1967).
15. CALLENDAR H. L. *Phil. Mag.* **32,** 104 (1891).
16. CAMPBELL C., GARDON S., SMITH C. L. *Anal. Chem.* **31,** 1188 (1958).
17. CARTER R. E. *J. Chem. Phys.* **34,** 2010 (1961).
18. CLAISSE F., EAST F., ABESQUE F. *Use of the Thermobalance in Analytical Chemistry,* Dept. of Mines, Province of Quebec, 1954.
19. COATS A. W., REDFERN J. P. *J. Polymer Sci.* **3,** 917 (1965).
20. COATS A. W., REDFERN J. P. *Analyst* **88,** 906 (1963).
21. COLLETTE G., JAQUÉ L. *Compt. Rend.* **251,** 2938 (1960).
22. CUEILLERON J., HARTMANSHEN O. *Bull. Soc. Chim. France* 172 (1959).
23. DEMASSIEUR N., MALARD C. *Compt. Rend.* **245,** 1514 (1957).
24. DOYLE D. C. *Techniques and Methods of Polymer Evaluation,* Vol. I., Dekker, New York, 1966, pp. 113–216.
25. DOYLE D. C. *Anal. Chem.* **33,** 77 (1961).
25a. DOYLE D. C. *J. Appl. Polymer Sci.* **5,** 285 (1961).
26. DUGLEUX P., DE SALLIER DUPIN A. *Bull. Soc. Chim. France* 973 (1967).
27. DUVAL C. *Inorganic Thermogravimetric Analysis,* 2nd Ed. Elsevier, London, 1963.
28. DUVAL C. *Mikrochim. Acta* 705 (1958).
29. EROFÈEV B. V. *Dokl. Akad. Nauk SSSR* **52,** 511 (1946).
30. EYRAUD C., GOTON R. *Bull. Soc. Chim. France* 1009 (1953).
31. FLYNN J. H., WALL L. A. *J. Polymer. Sci.* **3,** 917 (1965).
32. FLYNN J. H., WALL L. A. *J. Res. Natl. Bur. Stds.,* **70** A, 487 (1966).

33. FORSYTH R. C. *Proc. First Intern. Conf. Thermal Analysis, Aberdeen*, Macmillan, London, 1965, p. 246.
34. FREEMAN E. S., CARROLL B. *J. Phys. Chem.* **62**, 394 (1958).
35. FRUCHART R., MICHEL A. *Compt. Rend.* **246**, 1222 (1958).
36. GARN P. D. *Thermoanalytical Methods of Investigation*, Academic Press, New York, 1965.
37. GARN P. D. *Anal. Chem.* **33**, 1247 (1961).
38. GARN P. D., KESSLER J. E. *Anal. Chem.* **32**, 1900 (1960).
39. GARN P. D., KESSLER J. E. *Anal. Chem.* **32**, 1563 (1960).
40. GIBAUD M., GELOSA M. M. *Chim. Anal. (Paris)* **36**, 153 (1954).
41. GINSTLING A. M., BROUNSHTEIN B. I. *Zh. Prikl. Khim.* **29**, 1870 (1956).
42. GUIOCHON G. *Anal. Chem.* **33**, 1124 (1961).
43. HABER V., ROSICKÝ J., ŠKRAMOVSKÝ S. *Silikáty* **7**, 95 (1963).
44. HEGEDUS A. J., KISS B. A. *Mikrochim. Acta* 813 (1966).
45. HEGEDUS A. J. *Magy. Kem. Folyoirat* **72**, 79 (1966).
46. HEIDE K. *Naturwissenschaften* **53**, 550 (1966).
47. HLAVÁČ J. *Silikáty* **1**, 199 (1957).
48. CHATFIELD E. J. *J. Sci. Instr.* **44**, 649 (1967).
49. CHRISTIAN J. W. *Theory of Transformation in Metals and Alloys*, Pergamon, New York 1965, pp. 471—495.
50. *Instruments News*, Perkin Elmer Co. **18**, No. 2, 14 (1967).
51. JANDER W. *Z. Anorg. Allgem. Chem.* **1**, 163 (1927).
52. JACH J. *Phys. Chem. Solids* **24**, 63 (1963).
53. KEATTCH C. J. *Talanta* **11**, 543 (1964).
54. DE KEYSER W. L. *Nature* **172**, 364 (1953).
55. DE KEYSER W. L. *Bull. Soc. Franc. Ceram.* **20**, 1 (1953).
56. KNIGHT J. R., RHYS D. W. *The Platinum Metals in Thermometry*, Publication 2244, Engelhard Ind., Baker Pt Division, London 1961.
57. LAIDLER K. J. *Chemical Kinetics*, 2nd Ed., McGraw-Hill, New York 1965, pp. 15—17.
58. LE CHATELIER *J. Phys.* **6**, 23 (1887).
59. LUKASZEWSKI G. M., REDFERN J. P. *Lab. Pract.* **10**, 552 (1961).
60. LUKASZEWSKI G. M. *Nature* **194**, 959 (1962).
61. LUKASZEWSKI G. M., REDFERN J. P. *J. Chem. Soc.* 4802 (1962).
62. LUKASZEWSKI G. M., REDFERN J. P. *Lab. Pract.* **10**, 721 (1961).
63. MAGNUSON J. A. *Anal. Chem.* **36**, 1807 (1964).
64. MARCHE E. P., CARROLL B. *Rev. Sci. Instr.* **35**, 1486 (1964).
65. McADIE H. G. *Anal. Chem.* **39**, 593 (1967).
66. MARTINEZ E. *Am. Mineralogist* **46**, 901 (1961).
67. MAURAS H. *Bull. Soc. Chim. France* 260 (1960).
68. MIELENZ R. C., SCHIELTZ N. C., KING M. E. *Proc. Second Natl. Conf. Clays, Missouri*, 1953, Publ. No. 327, Nat. Acad. Sci., 1954.
69. MURRAY P., WHITE I. *Trans. Brit. Ceram. Soc.* **54**, 189 (1955).
70. MURRAY P., WHITE I. *Trans. Brit. Ceram. Soc.* **48**, 187 (1949).
71. NEWKIRK A. E. *Anal. Chem.* **32**, 1558 (1960).
72. NEWKIRK A. E., ALIFERIS I. *Anal. Chem.* **30**, 982 (1958).
73. NEWKIRK A. E. *J. Am. Chem. Soc.* **77**, 4521 (1955).
74. ERDEY L., PAULIK F., PAULIK J. *Acta Chim. Acad. Sci. Hung.* **10**, 61 (1956).
75. PAULIK F., PAULIK J., ERDEY L. *Z. Anal. Chem.* **160**, 241 (1958).

76. PAULIK F., PAULIK J., ERDEY L. *Acta Chim. Acad. Sci. Hung.* **26,** 143 (1961).
77. PAULIK F., PAULIK J., ERDEY L., LIPTAY G., GÁL S. *Periodica Polytechnica* **7,** 171 (1963).
78. PAULIK F., PAULIK J., ERDEY L. *Bergakademie* **12,** 413 (1960).
79. REICH L. *J. Appl. Polymer Sci.* **11,** 699 (1967).
80. RICHER A. *Inst. Rech. de la Sidérurgie* (*France*), Publication No. 187, Serie A., 1960
81. ROPP R. C., AIA M. A. *Anal. Chem.* **34,** 1288 (1962).
81a. RYNASIEWICZ T., FLAGG J. L. *Anal. Chem.* **26,** 1506 (1954).
82. SAITO H. *Sci. Repts. Tohoku Imp. Univ.* **16,** 1 (1927).
83. SIMONS E. L., NEWKIRK A. E. *Talanta* **10,** 1199 (1963).
84. SIMONS E. L., NEWKIRK A. E., ALIFERIS I. *Anal. Chem.* **29,** 48 (1957).
85. SIMONS E. L., NEWKIRK A. E. *Talanta* **11,** 549 (1964).
86. SERIN B., ELLICKSON R. T. *J. Chem. Phys.* **9,** 742 (1941).
87. SLADE P. E., JENKINS L. T. *Techniques and Methods of Polymer Evaluation*, Dekker, New York, 1966.
88. SHARP J. H., BRINDLEY G. W., NARAHARI A. B. N. *J. Am. Ceram. Soc.* **49,** 379 (1966).
89. ŠKRAMOVSKÝ S. *Chemie* **9,** 157 (1957).
90. SOULEN J. R., MOCKRIN I. *Anal. Chem.* **33,** 1909 (1961).
91. SOULEN J. R. *Anal. Chem.* **34,** 136 (1962).
92. ŠATAVA V. *Silikáty* **5,** 68 (1961).
93. ŠATAVA V., ŠESTÁK J. *Silikáty* **8,** 134 (1964).
94. ŠATAVA V. *Silikáty* **5,** 68 (1961).
95. ŠATAVA V., ŠKVÁRA F. *J. Am. Ceram. Soc.* **52,** 591 (1969).
96. ŠESTÁK J. *Dissertation*, Institute of Chemical Technology, Prague, Department of Silicates (1967).
97. ŠESTÁK J. *Silikáty* **7,** 125 (1962).
98. ŠESTÁK J. *Talanta* **13,** 567 (1966).
99. ŠESTÁK J. *Silikáty* **11,** 153 (1967).
100. ŠKRAMOVSKÝ S. *Chem. Listy* **26,** 521 (1932); *Collection Czech. Chem. Commun.* **5,** 6 (1933).
101. ŠKRAMOVSKÝ S. *Silikáty* **3,** 74 (1959).
102. TERREY D. R. *J. Sci. Instr.* **42,** 507 (1965).
103. TSUZUKI Y., NAGASAWA K. *J. Earth. Sci. Nagoya Univ.* **5,** 153 (1957).
104. VALLET P. *Anales de Chimie* **7,** 298 (1937).
105. VALLET P., RICHER A. *Bull. Soc. Chim. France* 148 (1953).
106. WATERS P. L. *J. Sci. Instr.* **35,** 41 (1958).
107. WATERS P. L. *Anal. Chem.* **32,** 852 (1960).
108. WILBURN F. W., THOMASSON C. V. *J. Soc. Glass Tech.* **42,** 158 (1958).
109. WILBURN F. W., HESFORD J. R. *J. Sci. Instr.* **40,** 91 (1963).
110. WILSON C. L. *J. Inst. Chem.* **83,** 550 (1959).
111. WADSWORTH M. E., HOLT J. B., CUTLER I. B. *J. Am. Ceram. Soc.* **45,** 153 (1962).
112. VALENSI G. *Compt. Rend.* **201,** 602 (1935).
113. VALENSI G. *J. Chim. Phys.* **47,** 489 (1950).
114. VALENSI G. *Compt. Rend.* **202,** 309 (1936).
115. WALKER R. F. *Vacuum Microbalance Technique*, Natl. Bur. Stds., Washington, Plenum Press, New York, 1962.
116. ZURAVLEV V. F., LESOKHIN I. G., TEMPELMAN R. G. *Zh. Prikl. Khim.* **21,** 887 (1948).
117. ZSAKÓ J. *J. Phys. Chem.* **72,** 2406 (1968).
118. HORÁK Z., KRUPKA F., ŠINDELÁŘ V. *Základy technické fysiky*, SNTL, Prague, 1961.

Chapter 3.

DIFFERENTIAL THERMAL ANALYSIS

3.1. GENERAL CHARACTERISTICS AND DEVELOPMENT OF THE METHOD

Differential thermal analysis (DTA) is a technique in which a record is made of the temperature difference between the sample and a reference material, against time or temperature, as the two specimens are subjected to identical temperature regimes in an environment heated or cooled at a controlled rate. The graphical record, the DTA curve, shows sharp increases or decreases in the temperature difference, depending on whether a change in the sample causes absorption or liberation of heat. The method records all changes in enthalpy, whether accompanied by a change in weight or not, e.g. phase transitions of first or second order (change of crystalline structure, boiling, sublimation, evaporation, melting), or chemical reactions such as redox reactions, decomposition, dehydration and dissociation. However, DTA indicates nothing about the *kind* of change taking place, either whether it is a phase change or a chemical reaction, or whether the change takes place in one or several steps. The nature and mechanism of the change can be analysed further by other methods (e.g. X-ray, and TG).

The method is closely related to the calorimetric method. However, in the latter the enthalpy changes are observed under static temperature conditions, while in DTA the temperature conditions are dynamic. A basic difference between DTA and many other thermoanalytical methods is the importance of the temperature difference between the sample and the programmed temperature. In thermogravimetry and dilatometry the properties of weight and volume are followed as a function of the temperature, which must be accurately controlled by the temperature programme. This is a prerequisite of accurate measurement. In DTA the existence of a difference between the temperature of the sample and that programmed (i.e. that of the reference sample) is, of course, the fundamental condition of the method. Thus, in TG, experimental conditions must be adjusted to eliminate the effects of self-heating or self-cooling of the sample as a result of the enthalpy changes occurring, while in DTA this self-heating or self-cooling is a necessary condition. In contrast to TG, the temperature of the sample in DTA is uncontrollable during the reaction and the resulting temperature effect cannot be directly related to the degree of reaction or physical change. This impairs the evaluation of kinetic data from DTA curves appreciably, as will be shown later.

The introduction of DTA is related to the discovery of the thermocouple. In 1878 Le Chatelier [80, 84] used thermocouples with photographic recording of the changes in e.m.f., in the study of the changes occurring in mineral substances on heating. This was done by following directly the temperature of the sample during heating or cooling. This simple method was improved upon by Roberts-Austen [128] who first modified the method to the differential form by using a reference thermocouple. The method was further developed mainly by improving the experimental arrangement and the methods of recording [14, 33, 56, 65, 74, 85, 114, 129, 134]. Although the fundamental idea and the arrangement of the measuring system are simple, proper exploitation of the method, mainly in quantitative analysis, is complicated owing to the number of experimental conditions which can fundamentally influence the results. These conditions, such as the basic geometrical arrangement of the sample holder and the furnace, the material used for the holder and crucible, the type of thermocouple and its contact with the sample, physical and chemical properties of the sample and the standard, method of heating and of temperature regulation, method of removing gaseous reaction products, the nature of the atmosphere in the reaction space, and the method of heat transfer, will be discussed in detail in the appropriate section. It has been observed that, as in TG, the results of various authors, obtained with different apparatus and under different experimental conditions, are often appreciably different, and that in DTA the results are more dependent on experimental conditions than they are in TG. This follows from the factors affecting the DTA curve, as the essence of the methods is the measurement of the temperature difference, which is much more dependent on the conditions and mechanism of heat transfer than are the weight changes which are measured in thermogravimetry.

Outstanding reviews of the literature were made by Grimm [53] and by Lehmann and co-workers [79]. Murphy [106−111] publishes biennial reviews of papers from the field of thermoanalysis, and independent surveys of the literature are published regularly [126]. Several basic monographs on DTA have appeared [47, 98, 99a, 123, 140, 145, 160, 166] which present both the theoretical principles of the method and a description of its development, as well as solutions to special problems, and possible applications. Zýka [167] has written on the applications of TG and DTA in analytical chemistry. Two interesting review articles [168, 169] give a good account of the use of DTA in the determination of heats of reactions and rate constants.

3.2. BASIC PRINCIPLES AND SCOPE OF THE METHOD

Every chemical reaction or physical transformation liberates or absorbs heat, causing a change in temperature. Such a change may also be accompanied by a change in weight corresponding to the formation of reaction products. DTA is capable of determining changes that are not accompanied by a change in weight, such as a change in crystal structure, or melting, which is its main advantage over thermogravimetry. Thus, DTA is the recording of every enthalpy change, exo- and endothermic, caused by any structural or chemical change. It should be kept in mind that DTA is a dynamic method in which equilibrium conditions are not attained; thus, the temperatures of changes and reactions determined in this way do not correspond to thermodynamic equilibrium temperatures. Generally, phase transitions, reductions, dehydrations, and some decomposition reactions are endothermic, while oxidation, some other decomposition reactions, and crystallization are exothermic. The position of the resultant temperature change on the temperature axis is characteristic of the investigated substance under the given experimental conditions, and may be used for its identification. The important factor for a given change is not the total amount of heat liberated or absorbed, but the rate at which this heat change, dQ/dt occurs. This determines the necessary sensitivity of the apparatus.

Energy changes which take place on heating or cooling the sample may be measured by various methods. The following are the most commonly used.

(a) Direct recording of the heating or cooling curve of the sample.

(b) Recording the inverse curve of the rate of heating.

(c) Recording of the DTA curve.

(a) DIRECT RECORDING OF THE HEATING OR COOLING CURVE, WHILE THE SAMPLE IS EXPOSED TO CONTINUOUS HEATING OR COOLING FOLLOWING A KNOWN LINEAR PROGRAMME. From Fig. 3.1 it is seen that the temperature function, $T = f(t)$, is linear up to the moment when the sample undergoes

change, its slope $f'(t)$ representing the rate of heating and being constant $[f'(t) = K]$. At the moment when an exo- or endothermic change takes place, the shape of the curve changes and the slope of the tangent assumes a different value. An exothermic effect causes an increase in the heating rate $[f'(t) > K]$, whereas an endothermic effect causes a decrease in heating rate $[f'(t) < K > 0]$. The main disadvantage in this recording method is its low sensitivity, because if a large temperature interval is to be covered, a relatively insensitive measuring device must be used and small temperature effects often cannot be detected. The method is suitable for recording relatively rapid phase changes at well-defined temperatures, and it is not very sensitive to transformations occurring over broader temperature ranges. It would be interesting to apply electronic differentiation to this type of temperature curve, thus obtaining the curve shown schematically in Fig. 3.1b. This curve could also be used quantitatively, and would have the advantage of eliminating the complicated measurement of the temperature difference between samples and reference (as in the classical DTA method). The author is not aware whether this method of recording the derivative curve is already employed in practice, or whether a mathematical treatment of such curves has already been given. The electronic differentiation of thermogravimetric curves has become common in commercial apparatus, and it may well be expected to find application in this case also. The possibilities and advantages of this approach have been discussed in the literature [169].

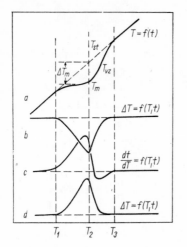

Fig. 3.1 Curves of the basic methods of thermal analysis

(a) Direct heating curve $\qquad T = f(t)$
(b) Derivative direct heating curve $\quad \Delta T = f(T, t)$
(c) Inverse rate curve $\qquad \mathrm{d}t/\mathrm{d}T = f(T, t)$
(d) DTA curve $\qquad \Delta T = f(T, t)$

T_1 is the temperature at the beginning of the effect
T_2 is the maximum difference of temperature
T_3 is the return of the temperature to the zero line.

(*b*) RECORDING THE INVERSE HEATING OR COOLING RATE CURVE. This consists of a continuous recording of the time required for an increase or decrease of sample temperature by a constant amount, such as 1 °C. This time is constant if the sample is heated at a linear rate and no enthalpy change occurs. A curve of this type, shown in Fig. 3.1c, is a straight line parallel to the horizontal axis, until the moment at which a heat change occurs, when the time interval either decreases (exothermic effect) or increases (endothermic effect). After the reaction the sample temperature returns to that of the surroundings; this is seen as a deviation in the opposite sense. This method is not widely used, mainly because it requires special recording apparatus, but it is much more sensitive than the method of direct recording. It has been described in papers [2, 35] including a discussion of possible quantitative applications.

(c) RECORDING THE DIFFERENTIAL CURVE, SHOWN IN FIG. 3.1d. The experimental procedure consists in heating or cooling both sample and reference material under identical conditions, and simultaneously recording the temperature difference between them. In an ideal case the heating or cooling takes place at a constant rate, the heat flow is identical in both materials and in the vessels used, the observed temperature difference is constant and the plot is horizontal. This line is called the base or zero line. The relationship $\Delta T = f(t)$ represents the heating rate relative to the reference sample. The first derivative, $d(\Delta T)/dT = f'(t)$ gives the relative acceleration or deceleration of heating rate that occurs when the sample undergoes a change which liberates or absorbs heat. This change is recorded as a deviation from the base (zero) line. If $f'(t) < 0$ the process is exothermic, if $f'(t) > 0$ it is endothermic. The temperature difference between the analysed and the reference samples is recorded by means of a so-called differential thermocouple, which is made from two thermocouples, one of which is placed in the sample and the other in the reference material, and connected in opposition. Some basic types of connection of the differential thermocouple are given in Fig. 3.2. In special cases thermistors or resistance thermometers can also be used, as will be shown later.

Every chemical reaction or physical change creates a peak on the DTA curve. Under suitable conditions, the temperature of the change taking place, as well as the heat of reaction and the rate of the process, may be determined. An idealized DTA curve is represented schematically in Fig. 3.3. From the initial temperature T_1 to temperature T_2 no reaction occurs. The difference in temperature between the thermocouples (ΔT) is zero. At T_2 the rate of the reaction increases and the endothermic effect becomes evident. The maximum temperature difference (ΔT) is reached at the point T_m. This point, however, is not important from the chemical point of view, because its position on the temperature axis depends on the thermal conductivity of the sample, the rate at which heat is absorbed as a result of the change taking place, and on the arrangement of the sample holders. The reaction proceeds up to the point T_x. At the point T_3 the sample has again reached the same temperature as the inert material ($\Delta T = 0$). In a similar way the subsequent exothermic effect can also be evaluated. By using this analysis of the DTA curve a fundamental qualitative and quantitative analytical idea of the observed process may easily be obtained. The temperatures of the origin and the peak of the effect (indicated usually by T_0 and T_m) are

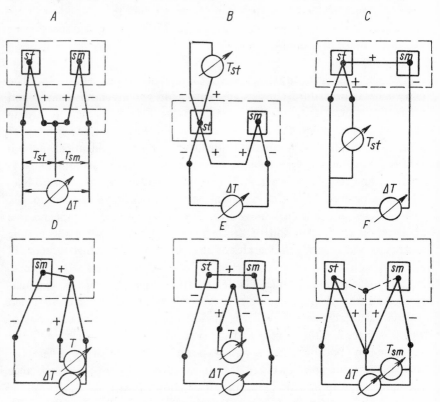

Fig. 3.2 Some basic methods of connecting the differential thermocouple

A. The temperature is measured both in the reference material and in the sample, the cold junctions are outside the heated space.

B. The temperature is measured in the reference material with an independent thermocouple, the cold junctions are outside the heated space.

C. The temperature is measured in the reference material, the junctions of both thermocouples are interconnected in the heated space.

D. The temperature is measured by the thermocouple junction located in the proximity of the crucible or the block, without the use of a standard.

E. The temperature is measured by an independent thermocouple located outside the samples, the cold junctions are outside the heated space.

F. The temperature is measured in the reference material, the cold junction is located in the heated space.

In cases *B* and *C* the temperature of the sample can be measured instead of that of the reference material if the containers are interchanged characteristic reaction temperatures, while the area between the curve from point T_1 to T_3 and the base line corresponds to the heat of reaction and is proportional to the mass of the sample. The shape of the curve may also give information on the kinetics of the reaction, as will be shown in the

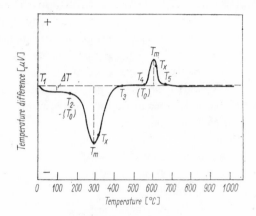

Fig. 3.3 Ideal DTA curve

T_1: Temperature of the beginning of the curve
T_2: Characteristic temperature of the beginning of the deviation of the curve from the zero line (usually indicated by T_0, as the temperature of the beginning of the effect).
T_m: Temperature of the peak.
T_x: Temperature of the end of the effect.
T_3 and T_5: Temperatures of the return of the curve to the zero line.
T_4: Temperature of the start of the exothermic effect.

appropriate section. As the DTA curve depends not only on the type of process under investigation, but also on the physical properties of the sample and on a number of experimental conditions, great attention should be paid to these parameters in the evaluation of the results.

DTA allows the detection of every physical or chemical change, whethre or not it is accompanied by a change in weight. The group of physical changes includes crystal modifications, phase transitions of a higher order, sublimation, evaporation, melting, crystallization, adsorption, and de-sorption. The group of chemical changes includes decomposition, oxidation, reduction, dehydration, reactions with the gas phase, reactions in the solid state, reactions in melts and chemisorption. In the case of a physical change of the first order a corresponding enthalpy change, ΔH, occurs at constant temperature, giving rise to an ideal DTA curve. This can be expected with reversible changes of crystal modifications, during melting and sublimation of solids (mainly metals). On cooling, an analogous course of the curve is obtained, but in the opposite sense. In practice, however, DTA of the majority of substances gives a far from ideal curve, mainly because of temperature gradients created in the samples. The determination of the actual temperature of the change (thermodynamic equilibrium temperature) is very difficult and it requires careful calibration of the temperature axis during the measurement of temperature directly in the sample, and a relatively slow rate of temperature increase, so that temperature gradients inside the sample decrease. The temperature at the beginning of the deviation from the zero line is thus closer to the thermodynamical equilibrium temperature than the temperature at the point T_m, in spite of the fact that it is also dependent on the properties of the apparatus, mainly the sensitivity. If supercooling does not occur, the equilibrium temperature may also be determined from the cooling curve. Supercooling or superheating leads to special effects on DTA curves, and is caused by the inhibition of the formation of nuclei of the new phase. Even in the simple case of melting, with some organic compounds small endothermic effects are observed at somewhat higher temperatures after the main endothermic effect is complete. This happens when the solid crystalline substance is transformed to a "crystalline liquid phase" during the main effect, and only the latter forms the melt at a higher temperature.

In the case of phase changes of a higher order there is no enthalpy change at constant temperature, but a discontinuous change in the specific heat and thermal conductivity is observed. This includes changes in magnetic properties, transitions from an ordered state to a disordered one, and devitrification. One of the uses of DTA is in the determination of phase diagrams of multicomponent systems on the basis of phase transitions as a function of the concentration of single components of the system. These transitions must be sufficiently rapid and free from inhibiting effects. The formation and decomposition of the mixed phases and compounds are, of course, observed in addition to the normal phase changes, and this usually requires further identification methods such as X-ray and infrared.

In chemical reactions it is mainly endothermic effects that are observed on DTA curves, usually from decomposition reactions and dehydrations. The reactions may be reversible and the equilibria are temperature- and pressure-dependent. Exothermic effects are met to a far lesser extent. They correspond to the liberation of energy during the decomposition of energy-rich substances, which are metastable at lower temperatures. The decomposition of explosive substances, and oxidation reactions, are examples. In comparison with physical changes, chemical reactions display a much more complex mechanism.

In thermogravimetry, the quantitative aspects are determination of the overall weight changes and the rate of the weight changes; in DTA they are the quantitative evaluation of the magnitude of the temperature effects and the shape of the curve. In an ideal case when both the weight and heat changes are measured under strictly equal conditions (e.g. when TG and DTA methods are carried out simultaneously), the temperature regions where changes occur will correspond. In the ideal case in TG the sum of all changes is directly proportional to the initial amount of the reacting substance, while in the case of DTA the relationship between the area under the curve and the heat of reaction or the amount of the active substance is not as simple, and must be determined individually. Reproducible quantitative results may only be obtained under suitable experimental conditions. Success in the application and interpretation of the results obtained by the DTA method is therefore closely connected with the instrumental technique and the practical procedure used. Usually information of a qualitative character can be obtained relatively easily, e.g. the temperature of the peak (T_m) usually characterizes the reaction observed, and the peak area is related to the corresponding heat of reaction and is approximately proportional to the amount of active substance. The shape of the curve may give valuable information about the reaction kinetics. If certain conditions arising from the theoretical basis of the method are not observed, the validity of the quantitative results is doubtful. The DTA method aims at the standardization necessary for obtaining reproducible results. Under such conditions it can be applied in the following fields.

(a) Qualitative and quantitative analysis of simple and complex materials and mixtures.

(b) Investigation of reaction kinetics and reaction mechanism.

(c) Determination of heats of reaction and specific heats.

The DTA apparatus is basically a dynamic calorimeter and its applicability is limited by its ability to detect thermal phenomena. A theoretical study of the basic processes determining the thermodynamic behaviour of the analysed system, i.e. heat-transfer mechanisms, kinetic processes connected with the change in physico-chemical properties and transport phenomena, is the basis from which the theoretical foundations of the method are deduced.

3.3. THEORETICAL BASIS OF THE DTA METHOD

As shown in the introduction, the major development of thermoanalytical methods has occurred only in the past twenty years. In this period, various mathematical procedures have been derived for the evaluation of results. Their introduction permitted quantitative evaluation of phenomena which earlier were only qualitative. Most of the earlier work was concentrated on elucidating causes of the poor correlation between the results of various authors and on quantitative applications of these results. The majority of these studies were carried out with mineral substances, which are rather complex materials. The introduction of a number of simplifications into the mathematical relationships in the analytical solution of the problem of heat transfer, as well as neglect of some thermodynamic aspects (the effect of atmosphere, pressure, etc.), sometimes considerably reduced the theoretical validity of the quantitative character of the method. It should be noted that a general and exhaustive theoretical solution of the DTA curve does not exist, because all theoretical considerations make use of certain simplifications. Only approximate relationships may be deduced in this way, expressing certain regularities of the DTA curve, but are very useful because they permit the expression of relationships between different experimental factors and the characteristics of the courves. The DTA curve may be characterized by use of a set of such properties of the recorded effects. Usually, when thermograms are evaluated, the position and magnitude of the temperature effect and the shape of the curve are used; these are determined from the temperature of the start, the peak, and the "area" of the effect. In fact, however the curve also has other, less specific, characteristics, such as the temperature at which the DTA curve deviates from the base line, the moment of formation of the quasi-stationary equilibrium, and a factor which expresses the shape of the curve from the point of view of reaction kinetics. These less specific characteristics will be discussed in greater detail in subsequent sections.

All theoretical considerations try to find a relationship between sample parameters and the area under the peak of the DTA curve. The physical and chemical phenomena are explained on the basis of the theory used for heat transfer and mass transport (diffusion). Both these transport mechanisms are inseparably connected, and they play a substantial part both in thermal kinetics, determined by the mechanism and the rate of heat transfer, and in the reaction kinetics. In the case of solid substances, heat transfer by conduction prevails and is therefore used in the majority of the theoretical equations deduced for the DTA curve. However, in complex systems in which a series of phase boundaries exists, and in systems containing powders (which is usually the case in DTA) the mechanism of heat transfer is more complex and therefore may not correspond to some of the simplifications applied to the mathematical solutions of the curve. The most difficult part in the solution of this problem is undoubtedly the question of the heat-transfer mechanism. For a more detailed study of this problem

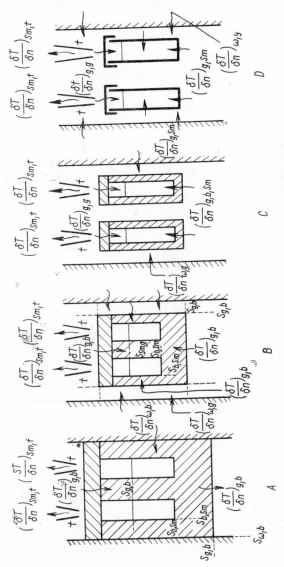

Fig. 3.4 Schematic representation of the cross-section of various type of differential thermal systems (sample holders) used in practice, with respect to the mechanism of the heat transfer. (With thermocouples).

A — close contact: B — system with medium contact: C — system with weak contact: D — free system Indices:
S — interface (boundary): n — normal in a given direction: g — gas: ω — furnace: sm — sample: b — block:
t — thermocouple

the literature [87−95] should be consulted. In addition to this, the situation is complicated by the fact that DTA has been applied in many specialized fields, which has led to the development of various types of apparatus, with detachable parts, such as sample carriers, crucibles, and the thermocouple system. This causes the mathematical solution of the curve to be directly dependent on the constructional method used, so that the simplifications introduced are not universally valid (particularly with respect to problems of heat transfer). The solutions are usually based on the simplifying assumptions that the heat transfer is by convection, that the physical constants of the sample do not change, and that the heating rate is linear. The requirement of convectional heat transfer is met in practice by using a sample holder in the shape of a compact block with two or more openings in which the sample and the standard are placed (see Fig. 3.4a, b). Some practical methods of heat transfer are represented schematically in Fig. 3.4. A development of increasing importance is that of applying DTA at sub-ambient temperatures. Suitable apparatus has been described [153].

A detailed derivation of the theoretical relationships used in the analyses of the DTA curve is beyond the scope of this book. Therefore, further consideration will be limited to the presentation of the final mathematical solution and some of the simplifying assumptions used. It may be considered that the simplest mathematical expression of the DTA curve should be based on analogy with liquids, where vigorous stirring can secure a uniform distribution of temperature in the sample, the standard, and their environment. This approach was taken by Borchardt and Daniels [27] and is shown schematically in Figs. 3.4 and 3.22. The authors introduced a number of simplifying assumptions of which the following should be mentioned: (a) a uniform distribution of temperature, (b) identical physical constants for both sample and reference (c_p, λ, ϱ) (c) the physical constants and the heat of reaction do not change within the temperature range of the reaction, (d) the temperature increase is linear. With these assumptions the changes in temperature of the standard and the sample are given by the equations

$$c_{p(st)} \cdot dT_{(st)} = (T - T_{(st)}) \cdot dt/a_{(st)} \tag{3.1}$$

$$c_{p(sm)} \cdot dT_{(sm)} = (T - T_{(sm)}) \cdot dt/a_{(sm)} = dQ_{(sm)} \tag{3.2}$$

where $a = \lambda/c_p \cdot \varrho$, λ being the thermal conductivity and ϱ the density.
Combining both equations gives

$$c_p \cdot d\Delta T + \Delta T dt/a = dQ \tag{3.3}$$

where ΔT is the temperature difference.

Integration of Eq. (3.3) gives a simple relationship expressing the heat of reaction:

$$c_p(\Delta T_\infty - \Delta T_0) + 1/a \int_0^\infty \Delta T \, dt = \Delta Q \tag{3.4}$$

where T_0 and T_∞ are the temperatures of the sample and the standard before and after the reaction. If they are equal, the first term of the left-hand side of Eq. (3.4) can be omitted and the integral of the second term expresses the area between the curve and the base line. This idealized model does not, of course, correspond to the actual conditions of measurement. There are other simplified approaches [25, 87−95, 122], but generally the solution starts with the basic equation describing the distribution of temperature in the given medium:

$$\varrho_{(i)} \cdot c_{p(i)} \cdot \left[\frac{\partial T}{\partial t} \right] = \lambda_i \left[\frac{\partial^2 T}{\partial x^2} + \frac{\partial^2 T}{\partial y^2} + \frac{\partial^2 T}{\partial z^2} \right] \tag{3.5}$$

where i indicates a given medium. The equation is then solved for one or two dimensions, usually for cylindrical or spherical symmetry.

One of the simplest expressions for the area under the DTA curve is the equation deduced by Speil and co-workers [143], and modified by Kerr and Kulp [64]. The final form of their equation is:

$$\frac{M_a \cdot \Delta H}{g \cdot \lambda_{sm}} = \int_a^c \Delta T \, dt \tag{3.6}$$

where M_a is the active mass of the reacting substance (grams), ΔH is the enthalpy change of the reaction per gram of active substance, g is a constant dealing with the effect of the geometry of the arrangement of the sample and of the standard on heat transfer, λ_{sm} is the coefficient of thermal conductivity of the sample, ΔT is the temperature difference, and a and c are the limits of the integral.

This expression relates the heat of the reaction ($M_a \Delta H$) to the area under the curve when constants g and λ_{sm} are introduced. The temperature drop in the sample and the dependence of the area on the specific heat of the sample are neglected. The expression may be evaluated graphically from the DTA curve by determining the area between the curve and a line drawn between the initial and the final inflexion points (area $T_2 T_m T_3$ in Fig. 3.3). Small differences (approx. 3 %) between calculated and measured values may be explained by the fact that the heat of reaction is not a linear function of temperature, but changes with temperature according to Kirchhoff's law:

$$\Delta H_T = \Delta H_0 + \int_0^T (a + bT + cT^2)\,dt = \Delta H_0 + aT + bT^2 + cT^3 \quad (3.7)$$

where ΔH_0 is the heat of reaction under standard conditions, and a, b, c are empirical coefficients. Equation (3.6) was obtained by direct integration of the original relationship deduced by Vold [154]:

$$\left(\frac{dy}{dt}\right) + A(y - y_s) = \frac{\Delta H}{C_{sm}}\left(\frac{df}{dt}\right) \quad (3.8)$$

where C_{sm} is the heat capacity of the sample and the vessel, f is the fraction of the active material of the sample reacting in time t, y is the temperature difference $(T_{st} - T_{sm})$, y_s is the temperature difference in the steady state attained after a sufficiently long time at initial conditions, and A is a constant.

At the beginning of the effect $y = 0$ and $t = 0$. The value of the temperature difference then increases to value y_s depending on the difference in heat capacity between the sample and the standard, the rate of heating, and the coefficient of heat transfer:

$$y_s = \frac{C_{sm} - C_{st}}{\lambda_{st} + \alpha_{st} + 2\dot{o}} \cdot \frac{dT_{sm}}{dt} \quad (3.9)$$

where C_{sm} is the heat capacity of the sample, C_{st} is the heat capacity of the standard, λ_{st} is the coefficient of thermal conductivity of the medium from the furnace wall to the standard material, α_{st} is the coefficient of thermal conductivity of the standard material to the surroundings, σ is the coefficient of thermal conductivity of the sample to the reference material, T_{st} is the temperature of the standard.

After the transformation or reaction, the temperature difference again attains the value $y = 0$ according to Eq. (3.8). The analysis of the DTA curve consists in plotting the value of $\ln(y - y_s)$ against time, starting from the top of the peak. After the end of the change the points fall on a straight line, thus indicating when the transformation or reaction has finished. The simplifying assumptions are constant heat capacity of the sample, and uniform temperature in the whole sample at any given moment. The heat capacity of the sample is given by the heat capacity of the crucible and of the reacted and unreacted part of the sample.

For blocks made of a material with high thermal conductivity (e.g. nickel) Soulé [144], Sewell [137], Eriksson [41], and Boersma [25] deduced equations to express the area under the curve. Boersma, for example, uses a nickel block (with infinitely large thermal conductivity) with two equal holes of cylindrical or spherical symmetry and deduces an expression for the area under the DTA curve. For a block with cylindrical holes, the area is given by the equation

$$\int_{t_1}^{t_2} \Delta T \, dt = \frac{q \cdot a^2}{4\lambda_{sm}} \tag{3.10}$$

where t_1 and t_2 are the times at the beginning and the end of the effect, q is the heat of transformation per unit volume, ΔT is the temperature difference, a is the radius of the hollow filled with the sample, λ_{sm} is the coefficient of thermal conductivity of the sample.

For a block with spherical cavities the following equation applies:

$$\int_{t_1}^{t_2} \Delta T \, dt = \frac{q \cdot a^2}{6\lambda_{sm}} \tag{3.11}$$

and for a plate, Eq. (3.12):

$$\int_{t_1}^{t_2} \Delta T \, dt = \frac{q \cdot a^2}{2\lambda_{sm}} \tag{3.12}$$

In the case of a ceramic block its thermal conductivity λ_c is of the same order as the thermal conductivity of the sample λ_{sm}. For an infinitely large ceramic block and extreme conditions $T = 0 = \int \Delta T \, dt$ and $r = \infty$, Boersma deduced the equation

$$\int_{t_1}^{t_2} \Delta T \, dt = -\frac{q \cdot V_{sm}}{\lambda_c} \cdot \int_{\infty}^{a} \frac{dr}{S} - \frac{q}{\lambda_{sm}} \int_{a}^{0} \frac{V_{sm}}{S} \, dr \tag{3.13}$$

where ΔT is the temperature difference, V_{sm} is the total sample volume, q is the heat of transformation per unit volume, qV_{sm} is the total heat liberated removed by the ceramic block, λ_{sm} is the coefficient of thermal conductivity of the sample, λ_c is the coefficient of thermal conductivity of the block, a the radius of the hole filled with the sample, r the radius of the block, and S the surface area of the sample.

In the case of an infinitely large ceramic block, Eq. (3.13) has no final solution for one- and two-dimensional cases. A solution exists only for a sample of spherical shape of radius a, where $V = (4\pi/3) \cdot r^3$, $V_{sm} = (4\pi/3) \cdot a^3$, $S = 4\pi a^2$. In this case the area is given by the equation

$$\int_{t_1}^{t_2} T \, dt = \frac{q \cdot a^2}{6} \left(\frac{2}{\lambda_c} + \frac{1}{\lambda_{sm}} \right) \tag{3.14}$$

If λ_c is infinite, the equation reduces to Eq. (3.11) for a spherical nickel block. Of course, in the case of a ceramic block the conductivities λ_{sm} and λ_c are of the same order and therefore the area according to Eq. (3.14) is larger than in the case of the nickel block. Although the ceramic block produces larger areas under the curve, a large number of samples in a single block may mutually interfere.

In classical DTA the sample produces heat, and by virtue of its thermal conductivity produces a measurable temperature difference. According to Boersma it is advantageous if these functions are separated by leading off the heat liberated by the reaction, through material in which the temperature difference may be measured. In this case the area is dependent only on the heat of the reaction and on the calibration factor of the apparatus, and does not depend on the volume or the thermal conductivity of the sample. The arrangement of the sample holder according to Boersma is shown in Fig. 5.1 in Chapter 5. The samples are placed in special nickel crucibles standing on a ceramic plate. During the reaction the liberated heat is conducted from the vessel through the ceramic plate which keeps the vessels at temperatures dependent only on the heat of reaction and the thermal conductivity between the vessels and the surrounding nickel covers. The greater part of the heat passes through the ceramic plate. The residue is transferred by convection and radiation which is accounted for in the calibration constant of the apparatus, and is independent of the type and the quantity of the sample. The area of the effect is then given by the equation

$$\int_{t_1}^{t_2} \Delta T \, dt = \frac{M_a \cdot q}{G} \qquad (3.15)$$

where M_a is the active mass of the sample, q the heat of reaction per unit of mass, and G the coefficient of thermal conductivity between the small vessel and the larger nickel cover.

Recently, the mathematical treatment of the DTA curve has been studied by Lukaszewski who submitted this method to a thorough critical analysis. In addition to the analysis of the method proper [87] he devoted himself to the deduction of equations for thermal conductivity in a given system [88], the setting up of physical and mathematical models [89], the initial and final conditions of heat conduction [90], the nature of heat transfer [91] and its mathematical treatment [92, 93] as well as the effect of the rate of heat transfer [94] and the basic conditions of the system under investigation [95]. In his extensive work he did not limit himself to a consideration of simple systems only, i.e. of compact metallic or ceramic blocks, but also considered systems where the block is not in direct contact with the furnace wall, or where both sample and reference stand freely and separately in the furnace. In such a case the connection between the calorimeter and the heat source is the atmosphere in the furnace. This is an arrangement often used to-day. From his work it is seen that the heat transfer in various thermoanalytical systems is complicated, and that the success of the theoretical solution of the DTA curve depends on strict testing of each model and the results to which it leads. The connection between heat transfer and kinetic processes is determined by some physical parameters and by the properties of the given system in which the process takes place. The properties of the system cannot be altered by changes in the experimental set-up.

The application of mathematical theory to DTA contributed to the quantitative elucidation of some of the results obtained in the course of the extensive development of the method. Most of the mathematics derived so far does not deal with basic aspects of heat transfer or the experimental technique used, but is directed towards quantitative evaluation of both the amount of the material investigated, and the corresponding heat of reaction (which lies in the field of calorimetric methods). However, even with these solutions a number of complex problems still remain unsolved. They may be dealt with by the introduction of some simplifications into the mathematical considerations; however, these do not often correspond accurately to the physical and chemical basis of the problem. To explain these relationships it is necessary to express mathematically the mechanisms of heat transfer and mass transport at their various stages. The papers by Lukaszewski are aimed in this direction. Both transport mechanisms are inseparably connected, and they are important not only because of their thermal conductivity aspect, determined by the rate of heat transfer, but also because of the reaction kinetics aspects governed by basic thermodynamics. The quantitative treatment of the DTA curve may lead either to the determination of the amount of the active substance, i.e. quantitative analysis, or to the determination of specific heats or heats of reaction, or else to the evaluation of kinetic constants. The evaluation of the curves from the point or view of quantitative analysis, heats of reaction and kinetic constants, will be discussed in the appropriate sections.

In conclusion, it may be said that a series of mathematical expressions for the DTA curve have been deduced, based on general physical laws of heat and mass transport. However, in view of the possible arrangements of the system the solution proper is complex and is not universally valid. Experience with practical evaluations of the curves shows that the most convenient method, expressing the relationship between the area limited by the curve and the heat of reaction, is an empirical one.

3.4. BASIC FACTORS AFFECTING THE DTA CURVE

The main quantities which are determined from a DTA curve are the following: temperature of the beginning and the end of the temperature effect, temperature of the peak maximum, area under the curve, and in quantitative evalutions, the determination of the amount of active material or the corresponding heat of the transformation. As in thermogravimetry, poor agreement between the results obtained with various apparatus and by various authors has been observed, mainly with respect to the value and shape of the peak maxima. Therefore, great attention has been devoted to standardization of the method itself, and to the determination of factors which might possibly affect the curve. It has been found that the results of DTA are appreciably affected by factors connected with the type of apparatus, physical and chemical nature of the investigated substance, and also with the technique employed. Hence, although DTA is a relatively simple method, sufficiently sensitive for the detection of very small energy changes, and capable of detecting even small changes in a relatively broad temperature range, great attention must be devoted to the instrumentation and method, as well as to the physical and chemical properties of the substances investigated. These factors were studied by Arens [3], primarily with mineral substances. The factors which may affect the DTA curve and determine the choice of the experimental conditions, may be divided into three main groups:

Instrumental characteristics

(*a*) Heat source and temperature regulation system. (Method of heating, shape and size of furnace.)
(*b*) Detection system for the DTA curve. (Construction of the calorimetric part itself, material of the block of the detecting system, geometry of the block, size of holes or crucible, etc.)
(*c*) Systems of temperature measurement. (Kind and size of thermocouple junction, its position in the sample and the fixing of thermocouples.)
(*d*) System of DTA curve recording. (Rate and sensitivity of the recording system.)

Method

(*a*) Nature of the sample. (Size of the sample, pretreatment, packing, etc.)

(b) Effect of atmosphere and the type of contact between the atmosphere and the sample. (Effect of pressure and composition.)

(c) Type of heating. (Rate, linearity and regulation.)

Sample and reference material

(a) Physical and chemical properties of the sample.

(b) Physical and chemical properties of the reference material.

(c) Dilution of sample.

The most important of these factors are the size of particles, degree of crystallization, thermal conductivity, density, heat capacity, and the effect of diluting substances.

Some of these factors are interdependent and cannot be considered separately. Under certain conditions many of them may be controlled, e.g. the linearity of the heating, regulation of the conditions of the heat flow. Other factors and their changes in the course of the experiment are controllable only with difficulty or not at all, e.g. changes in heat capacity or thermal conductivity, which lead to changes in the baseline. These factors will now be discussed in greater detail.

3.4.1. Instrumental characteristics

(a) Heat source and temperature regulating system

These two factors determine in principle the method, the rate, and the linearity of heating. The type of heating is one of the most important factors, affecting both the position of the temperature peak (T_m), and the area under the curve. In current practice, heating rates range from 0.1 °C/min up to 200 °C/min but for normal work rates of $8-12$ °C/min are usually used.

Measurements have shown that a higher rate of heating leads to a shift in T_m towards higher temperatures and increases the area under the curve to a certain extent. Although at higher heating rates a larger proportion of the reaction takes place in the same time interval, leading to an increase in the rate of enthalpy changes of the reaction ($\Delta H/dt$), the effect depends on the type of reaction taking place. From Eq. (3.4) it follows that the temperature of the peak is characterized by the temperature difference and the rate of heat absorption. A higher rate of heating causes an increase in the value $\Delta H/dt$, and hence also of $(\Delta T)_m$. As the return to the zero line is a function of time and temperature difference, this return takes place at a higher temperature when the heating rate is increased and also depends on the type of reaction. In a case where a sample undergoes only a phase change without a change in weight, the effect of the heating rate is much less than in the case of a chemical reaction accompanied by a change in weight. In the second case a larger shift occurs when the rate of heating is increased. This may be explained by assuming that the gas liberated in these reactions (e.g. water vapour in dehydrations) has a tendency to prevent further decomposition. The diffusion of the gas away from the reaction vessel is time-dependent and the increased heating rate may affect the partial pressure of the gas in the vicinity of the sample, and thus

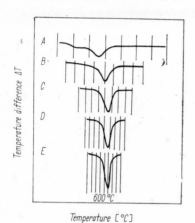

Fig. 3.5 Effect of rate of heating on the DTA curve of kaolin. Rate of heating: A — 5 °C/min, B — 8 °C/min, C — 12 °C/min, D — 16 °C/min, E — 20 °C/min.

also the course of further decomposition. This is true primarily in dehydration reactions, where — as in the case of kaolin — a shift of T_m towards higher temperatures was observed (see Fig. 3.5) when the rate of heating was increased [3, 68, 142]. A similar effect was also observed during the decomposition of carbonates.

On the other hand, with slow heating the decomposition product (e.g. water vapour) diffuses into the atmosphere more readily and its partial pressure does not attain a high value. Therefore, the decrease in temperature differences on slower heating is due not only to the fact that the sample is closer to thermal equilibrium as a result of heat transfer, but also to the fact that as a result of the diffusion of the gaseous products, the temperature of the decomposition also decreases. From Fig. 3.5 it is seen that the size of the effect increases as the heating rate increases. This was shown theoretically [68, 123] by deduction of the expression:

$$\ln (\Delta T)_m = A' - \frac{E}{2.303 R T_m} \qquad (3.16)$$

where A' is a constant independent of the peak temperature, and T_m is the temperature of the peak. According to this relationship, $\ln (\Delta T)_m$ is a linear function of $1/T_m$. This was verified experimentally [3].

Another effect of an increase in the heating rate is a decreased resolution of reactions which follow rapidly one after the other, e.g. a series of dehydrations, where the steps which at a slow heating rate are distinctly separated, coalesce to a single step at an increased heating rate.

The rate of heating may also substantially affect the whole DTA curve by the fact that at high heating rates the amount of gaseous reaction product alters the composition of the furnace atmosphere and thus also the further course of the reaction. This effect was observed by Kissinger [68] in the case of $FeCO_3$, and Blažek [18] in the case of $MnCO_3$ (see Fig. 3.6).

As a result of the increase in the heating rate a rapid evolution of CO_2 prevents the access of oxygen from the atmosphere to the sample, thus altering the exothermic effect of oxidation. It should be noted that the shape and the reproducibility of the curves obtained will depend predominantly on the amount of the sample as well as on the shape and the size of the crucible and the reaction space. The heating rate also sometimes affects the magnitude of the area under the curve used in quantitative evalua-

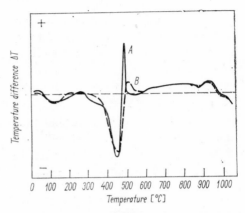

Fig. 3.6 DTA curves of $MnCO_3$ in air at various heating rates. A — 3 °C/min, B — 10 °C/min. (Reprinted from [18] by permission of the copyright holder)

tion. The results of various authors are somewhat different in this respect, and their explanations are complicated. In principle this effect concerns all types of transformations, even those that do not involve a weight change. The effect is probably connected with heat-transfer mechanisms. Proks [125] studied the effect of the heating rate on the shape and the size of the area under the curve and its position on the temperature axis. He [125] and Piloyan [123] deduced approximate mathematical relations between the constant rate of heating and the quantities important for the evaluation of DTA curves, i.e. the magnitude and the area under the curve, maximum value of the temperature difference between the standard and the sample, and the temperature interval within which the investigated reaction takes place. They came to the conclusion that for quantitative determination of the area, recording the dependence of ΔT on time is better in view of the difficulties connected with accurate maintenance of a constant heating rate of the standard, while the record of the dependence of ΔT on the temperature of the standard is better when the temperature of the investigated reactions is determined. An expression for the area, B, was deduced by a series of authors [3, 25, 125, 142]:

$$B = \frac{\bar{Q}\varrho \cdot r^2}{4\bar{\lambda}} \qquad (3.17)$$

where $\bar{Q}$ is the heat liberated during the change per unit mass of the starting material, ϱ is the density of the starting sample, $\bar{\lambda}$ is the mean value of the thermal conductivity, r is the radius (if cylindrical or spherical symmetry is considered).

From this equation it follows that the area under the curve is proportional to the amount of the sample, and to the size of the vessel, and that during quantitative analysis the dependence of B on peak temperature should be taken into consideration. Barrall and co-workers [10] and Vold [154] used extrapolation to zero rate of heating in the determination of the melting points of some organic compounds. As will be shown in section 3.8, T_m depends on the rate of heating and this dependence may be used for the study of reaction kinetics [68, 69].

The effect of the heating rate on the character of the DTA curve may be summarized according to Arens, as follows:

(a) A substantial difference exists between reactions accompanied by change in weight (i.e. chemical reactions) and reactions with no change in weight (phase changes).

(b) The reactions accompanied by weight changes are strongly affected by the heating rate. With increasing heating rate the maximum temperatures of peaks increase, as do heights and areas. However, the width of the curve, corresponding to the time of the reaction, decreases.

(c) If the temperature of the furnace is measured directly in the sample, then the transformations during which no weight changes take place (phase transitions) do not affect the temperature of the peak T_m, but they do influence the amplitude and area under the curve.

(d) From the results obtained at various heating rates, while other parameters remain constant, it is possible to calculate heats of reactions for a zero rate of heating (for example static dehydration) and to study the correlation between DTA and dehydration studies.

It is evident that the definition of the position of the temperature effect on the temperature axis is of great importance. This position can be best characterized by points corresponding to the beginning of the effect (T_0) and the peak (T_m). The standardization committee of ICTA also recommend additional evaluation of the half-width of the peak.

(b) The detection system

The construction of the block i.e. the specimen holder and the material used are other basic factors affecting the curve. Both appreciably affect the shape of the curve and the magnitude of the effects.

At the present time copper, silver, nickel, chromium-nickel steel, platinum, and ceramic materials, mainly sintered corundum, are commonly used for making blocks. As was explained earlier the main purpose of the block is to transfer heat to the sample. This depends on the thermal conductivity and heat capacity of the material used. Emissivity is also important, especially at temperatures at which heat transfer is by radiation. As polished metallic surfaces have an emissivity of $0.1-0.25$ in the temperature range $20-1000$ °C (compared with 1.0 for ceramic material), this property may be decisive in choosing a suitable material. At these temperatures, heat transfer may therefore be more rapid in a ceramic block. From the large amount of data on the comparison of ceramic and metallic blocks, the conclusion can be drawn that a block made of material with low thermal conductivity (ceramic) produces a better resolution of endothermic effects than a block made of a good conducting material (for example nickel), and in the case of exothermic reactions the resolution is better with metallic blocks [3, 47, 52, 98, 158, 160]. Ceramic blocks seem more sensitive at lower temperatures, metallic blocks at high. The reason for this is the mechanism of the heat transfer in which the emissivity of both materials is involved. At lower temperatures the relatively weakly thermally conducting ceramic block insulates both sample and reference sufficiently, while in the range of radiation mechanism the high emissivity of the ceramic material accelerates the heat transfer. In a similar way, at higher temperatures the properties of the metallic block come into effect. Another temperature effect may be due to inefficient heat transfer between the sample and the block. The heat transfer in a system making use of a block, i.e. heat transfer between the furnace wall and the sample within the block, depends on the material of the block, its emissivity, thermal conductivity, and thermal diffusivity, defined as

$$a = \frac{\lambda}{\varrho \cdot c} \tag{3.18}$$

where a is the thermal diffusivity, λ is the thermal conductivity, ϱ is the density, and c is the specific heat. In comparison with metallic materials (e.g. nickel), ceramic materials possess a higher emissivity, and, other conditions being equal, this allows more rapid heating of its surface. In contrast, the rate of heat transfer inside the block is controlled by its thermal conductivity, which is lower than that of a metallic block.

With respect to this, an interesting experiment was conducted by Garn [47] who heated two geometrically identical and relatively thin-walled blocks under identical conditions, one made of nickel and the other of sintered corundum, and measured the temperature difference between the thermocouples placed in the centre of Al_2O_3 samples. He found that in almost the whole temperature range investigated the corundum transferred heat to the sample more rapidly, and that the nickel was more efficient only above the temperature of its Curie point, where the continuous decrease of its thermal conductivity changes abruptly. It would be interesting to carry out a similar experiment with larger blocks as well. The plot of ΔT for this arrangement is shown in Fig. 3.7. Other complications in heat transfer may be caused by various surface effects such as corrosive coatings.

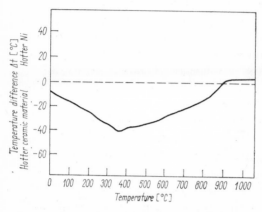

Fig. 3.7 Temperature difference between alumina samples enclosed in ceramic and nickel blocks versus the temperature of one of the samples. The ceramic transfers heat more rapidly for the greater part of the temperature range.
(Reprinted from [47] by permission of the copyright holder)

Heat transfer, and thus the shape of the DTA curve, is also affected by the porosity of the block, on which depends the diffusion of the gaseous product from the sample. This is the reason why in the case of reactions liberating gaseous products the temperature effects are shifted to lower temperatures when ceramic blocks are used [158]. From a comparison of curves obtained with nickel and ceramic blocks it is seen that with ceramic blocks the endothermic effects are greater and the whole area is usually approximately $20-30\%$ larger than with metallic blocks. It also seems that with a metallic block the endothermic reaction starts at a higher temperature. This can be explained by the fact that in the case of a metallic block with high thermal conductivity a rapid heat flow into the surface layer of the sample neutralizes the initial endothermic effect. Hence, it seems that a ceramic block should often be more advantageous, especially for the observation of endothermic rections, because the effects obtained are larger.

On the other hand, a number of samples (in the same ceramic block) may interfere with one another [25].

The next problem which should be studied in estimating the effect of the block is the shape and size of the cavities for the samples. In this respect a series of modifications and shapes has been proposed, ranging from massive blocks to a more or less crucible type modification, arranged both horizontally and vertically. The vertical arrangement is prevalent to-day, because symmetry of the arrangement is more easily obtained, which is necessary for thermal equilibrium of the differential thermocouple. The construction of the block will be shown in Chapter 5. Here, attention will be devoted mainly to the influence of the geometry of the block on the curves. A substantial difference occurs when a massive block is used instead of a crucible type modification (which is met more often, especially in recent years when the detection of temperature differences from the crucible without the use of the block, which is similar to a calorimetric arrangement, is more frequent). The crucible modification is often used when the DTA curve is recorded simultaneously with the thermogravimetric curve [17, 104, 121]. It has been found that sample holders in the form of isolated crucibles give a greater sensitivity than the blocks commonly used. They also display a greater temperature inhomogeneity, which might affect the course of the temperature effect adversely. This danger may be limited to a certain extent by using thick-walled crucibles.

The quantity of the analysed sample, the construction of the block and the size of the cavities for the samples have a substantial effect on the shape and magnitude of the temperature effects. The smaller the diameter of the sample, the smaller will be the temperature gradient in the sample, and thus the peak will be formed in a shorter time interval and be smaller. A smaller amount of sample in the immediate proximity of the thermocouple means also that the danger of the zero line being distorted is lower. The most commonly used shape for the sample is cylindrical, which allows a symmetrical disposition of the thermocouple. From the theoretical point of view the most suitable shape is spherical, as in this case heat reaches the centrally placed thermocouple junction from all sides through the same phase. However, realization of such a situation is difficult and Lehmann [79] proposed the use of a cylindrical cavity with a spherical bottom. Diameters of the cavities or of the sample crucibles currently used are approximately 12 mm or less. The crucibles are of various shapes, usually with a concave bottom, with the thermocouple junction introduced from below. Miniature crucibles, and crucibles placed inside the thermocouple junctions are also known (pp. 149, 229). At a given heating rate the area under the curve increases with the size of the cavity. With an increasing quantity of sample, the peak shifts to higher temperatures because the time required for the reaction is longer. The magnitude of the shift is dependent on the sample under investigation and the reaction rate. With an increasing amount of sample the width of the peak increases, and hence resolution of neighbouring effects is poorer. The geometry of the sample carrier may also influence appreciably the diffusion of gases from the sample as was shown by Mackenzie [98] and Webb [158]. Enclosing the sample with a lid shifted the peaks to appreciably higher temperatures and led to distortion as a result of the formation of a static atmosphere. The shape and size of the furnace also depends on the geometry and the size of the blocks, as does its position with respect to the block. In this respect smaller internal diameters of the furnace (approx. $25-37$ mm) are preferred to-day, because of the formation of a less steep temperature gradient. The heat transfer depends directly on the geometrical arrangement of the relevant parts of the apparatus. Figure 3.4 shows the basic possible alternatives. From the point of view of heat transfer, D is the simplest; it makes use of separated crucibles, or, as will be shown later, in the case of a special micro-arrangement, only thermocouple junctions with suitable cavities. A simplification may thus be introduced into the basic equation for heat transfer (3.5) by taking only convective heat transfer into account. The use of compact blocks with cavities for the samples corresponds to this, but decreases the total sensitivity of the apparatus.

Another factor related directly to the factors involved in the block and its geometry is the effect of the quantity of sample. It follows from theoretical considerations that the area of the peak should be directly proportional to the amount of the active substance. A number of papers have been published investigating this relationship in detail [5, 10, 34, 162] and concluding that the relationship is only valid in certain concentration ranges. However, the results cannot be generalized because they are dependent on the apparatus used, but it can be said that deviations from linearity only occur after certain concentrations are exceeded, as a result of the reaction equilibrium being reached more slowly with larger samples.

In conclusion, the effects caused by the geometry and the material of the block may be summarized as follows [3].

(a) A ceramic block produces curves which are well-shaped for endothermic effects and poorer for exothermic effects.

(b) A metallic block produces narrower peaks, endothermic peaks being less well-shaped than exothermic ones.

(c) The effect of the block is apparent in all types of reactions whether a change in weight of the sample occurs or not, i.e. whether the process is a phase change or a chemical reaction.

(*d*) The size and shape of the sample cavity affect primarily reactions accompanied by a change in weight, mainly with respect to the magnitude of the effect, its temperature and the time of the reaction. Reactions which are not accompanied by a weight change, e.g. changes in crystal modification, are affected only with respect to the magnitude of the temperature effect.

(c) Temperature measurement and effect of the thermocouples

The DTA curve is affected by the thermocouples in various ways, the most important being the effects due to the position, type and size of the thermocouple.

Each of these effects may be considered in various ways, e.g. the effect of the position of the thermocouple may be considered in terms of temperature measurement for graduation of the temperature axis, or of the effect on the magnitude of the temperature effect, or on the position of characteristic points on the curve, or correspondence of these points to the actual temperature at which the change takes place. The effect of the type and the size of the thermocouple is studied mainly from the point of view of thermocouple materials, size of the junction, and the strength of the thermocouple wires. Generally, these effects may be divided into two groups, viz. effects arising from the position of the thermocouple junction in the sample and the standard, the position used for measurement of the programmed temperature, and the effect due to the material and the mechanical strength of the thermocouple used. The position of the thermocouple junctions inside the sample may affect not only the position of the peak on the temperature axis, but also its shape and magnitude. The method of temperature measurement for recording the temperature axis may appreciably influence the result of the analysis. For practical purposes the temperature is usually measured at the centre of a cylindrical sample, because this is where the reaction occurs last, as a result of heat transfer from the cavity wall to the centre. Therefore, the temperature difference ΔT is greatest here, and the temperature of the peak corresponds in this case to the temperature of the reaction. When the temperature of the sample is recorded as a function of time, no complications arise in determining the characteristic points of the curve, because the temperature is recorded sinultaneously. However, from the DTA curve no conclusions can be made regarding the temperature of the sample or the reference. In order to do this, the temperature should be measured by another thermocouple (or by a combined differential thermocouple). The position of its junction in the sample, the reference material, or the block is of great importance for the correct determination of characteristic points on the temperature axis. Views on the best position for the thermocouple used for the recording of temperature axis differ greatly. In practice the thermocouple is often placed in the reference material or the block, but seldom in the sample. However, it should be borne in mind that the determination of the correct temperature of the change is in no way simple, because of the existence of a temperature gradient inside the sample, which depends not only on the properties of the material itself, but also on the rate of heating, and because at the moment the reaction occurs a discontinuous change takes place. If the temperature gradient in the reference material is known for a given heating rate and temperature and it is assumed that the sample behaves similarly, a correction can be applied to the sum of the temperature difference and the measured temperature of the block to obtain the sample temperature. Some authors argue that when the thermocouple is placed in the analysed sample, DTA curves of the same sample measured on different apparatus at various heating rates cannot be compared, and they recommend placing the thermocouple in the reference material or the block to enable this comparison to be made. Of course, in this case the true

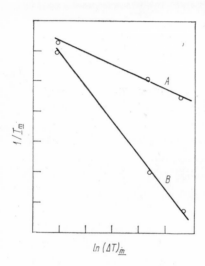

Fig. 3.8 Plot of $1/T_m$ vs. $\ln(\Delta Z)_m$.
A — temperature measured in the sample; B — temperature measured in the reference material
(Ref. [123], p. 181)

temperature of the sample is known. As in the thermogravimetric method, this knowledge is very important in some measurements, e.g. in the determination of kinetic constants. The problem consists of the fact that errors may be made in the determination of the temperature of characteristic points, and that the temperature of these points is dependent on the heating rate in a different way from the temperature measured in the standard. This is evident from the plots of $1/T_m$ versus $\ln (\Delta T)_m$, Fig. 3.8, (where T_m is the peak temperature, in K) for identical samples of kaolin, obtained by measuring the temperature in the sample (curve A), and in the reference material (curve B) [123]. In both cases the plots are linear, but the activation energies determined from their slopes are different, viz. 40 kcal/mole in the first case, and 11 kcal/mole in the second. The first value agrees well with the results of other measurement [68]. From this it is evident that the determination of the true temperature of the sample is very important and that a simultaneous recording of the relationships between ΔT and t, T_{sm} and t, and T_{st} and t may supply valuable information.

A further effect on the curve may be due to an incorrect position of the differential thermocouple, which can lead to various deformations of the curve. The position producing the maximum and well-defined peak is that where the thermocouple junctions are fixed symmetrically in the centre of cylindrical samples. The problem of the position of the thermo-couples was studied by Smyth [141], who observed the distribution of the temperature in the reference and the analysed sample, and who deduced the theoretical shapes of the DTA curves as a function of the place where the temperature was measured. The distribution of temperature in the reference sample is given by the equation

$$T = T_c + \alpha t + \alpha x^2/2a \qquad (3.19)$$

where T is temperature, t is time, x is distance along the direction of the heat flow, α is the rate of heating, a is thermal diffusivity and T_c is a constant.

This equation shows that at every instant the temperature distribution is characterized by a parabola. The temperature distribution curve in the analysed sample is given, initially, by the same parabola, but when the point when the reaction starts is reached, the flow of heat into the centre of the sample is reduced because it is absorbed in the reaction, thus causing the deviation of the curve from the baseline. At this moment the temperature inside the sample does not correspond to the temperature of the transformation. The completion of the reaction throughout the sample results in a sharp temperature reversal (an increase) according to the equation of heat flow:

$$\frac{\partial^2 T}{\partial x^2} = \frac{1}{a}\frac{\partial T}{\partial t} \qquad (3.20)$$

where x is the distance along the direction of flow, T the temperature, t the time, and a the diffusivity. This reversal decreases gradually until the sample temperature is again equal to the temperature of the standard. Therefore the shape of the curve and the temperature of the peak T_m are a function of the position of the reference thermocouple. Smyth compares theoretically deduced curves obtained by plotting the temperature difference against the temperature of the centre of the reference sample, and thus clearly illustrates the differences. The temperature is the temperature

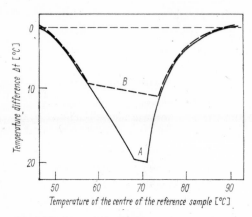

of the reaction only when measured at the centre of the sample being analysed. Smyth's results can also be used to show the effect of an asymmetrical position of the thermocouple junction, which leads to curve deformation of various magnitudes [10, 141]; see Fig. 3.9.

Fig. 3.9 Deformation of DTA curves as a thermocouple junction is moved from a symmetrical position (the further it is moved the greater the deformation) (*A* — thermocouple located 0.06 cm from the sample center; *B* — thermocouple located 0.3 cm from the sample center) (Reprinted from [10] and [141] by permission of the copyright holders)

This deformation is evident mainly as a truncation of the peak. The effect increases as the thermocouple junction is moved further from the symmetrical position, which increases the time during which the thermocouple is in contact with the new, high-temperature form, before a complete change is achieved. An asymmetric position of the junction of the differential thermocouple, as well as asymmetric heating, leads to the formation of distorted curves. It was found that with a symmetrically placed thermocouple, the T_m value increased with the heating rate. The effect of the position of the differential thermocouple junction on the shape of the DTA curve was also studied by Jankowski [61] who used a cylindrical vessel and placed the thermocouple junction at various levels in the axis and near the walls of the vessel,

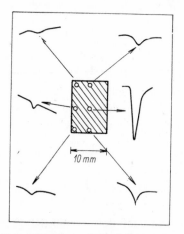

Fig. 3.10 Graphic illustration of the size of thermal effect due to $\alpha \rightleftharpoons \beta$ transformation of SiO_2 showing the effect of different thermocouple positions. The cross-section of the sample is drawn in the middle of the figure (reprinted from [61] with the permission of the copyright holder)

as is shown in Fig. 3.10. The greatest value and optimum form of the peak were obtained when the junction was in the centre of the sample.

Figure 3.11 shows that a proportional increase of the effect cannot be achieved simply by increasing the amount of sample [170].

It is evident that the position of the thermocouple and the method of fixing the temperature axis are of great importance in the evaluation of thermograms. The conclusions from the preceding considerations may be summarized as follows.

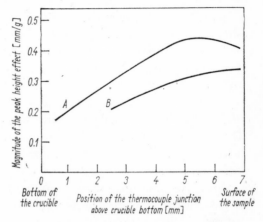

Fig. 3.11 Size and shape of thermal effects as a function of the position of the thermocouple junction in a SiO$_2$ sample.
A: junction in the axis of the sample
B: junction near the wall of the crucible [170].

(a) If the temperature difference, ΔT, is expressed as a function of the temperature of the surface of the sample, then the temperature of the point at which the curve first deviates corresponds to the temperature of the change.

(b) If the temperature difference, ΔT, is expressed as a function of the temperature of the sample centre, and if the temperature of the sample surface increases linearly, then the temperature of the peak corresponds to the temperature of the change.

(c) If the temperature difference, ΔT, is expressed as a function of the temperature of the centre of the reference material, then neither the point of the first deviation of the curve nor the temperature of the peak corresponds to the temperature of the change.

(d) If the peak temperature is measured inside the sample then it does not depend on the heating rate in the range 5−80 °C/min.

(e) If the temperature is measured in the reference, then the point of the first deviation of the curve depends on the heating rate.

(f) If the change takes place continuously within a temperature interval, then it cannot be characterized by a single temperature. It may be characterized, however, by the temperature at which the rate of change under the given conditions attains its maximum value.

The type and size of the thermocouple are just as important as its position. In the literature, thermocouples of dimensions varying from

0.08 to 0.5 mm are described. As the thermocouple is heated simultaneously with the sample, its size will be important because of its thermal conductivity and heat capacity. This problem was treated theoretically by Boersma [25] and Sewell [137] who showed that part of the heat liberated from the sample is removed by the thermocouple lead and thus the area of the peak is diminished. Boersma considered spherical and cylindrical samples in a nickel block with a thermocouple junction situated exactly symmetrically, and deduced theoretical relationships for the amount of heat removed by the thermocouple. According to these authors the removal of the heat may cause a decrease in the peak area of up to 40% of the theoretical value. The area is also sensitive to the geometry of the thermocouple. When the thermocouples were changed, changes of up to 20% in the calibration factor were observed. The dimensions of the thermocouple and of its junction may also affect the area under the curve [55, 153].

Thermocouples are usually used for the measurement of temperature and temperature difference, but if in a special case high sensitivity is required, thermistors may also be used [118, 157]. However, they limit the temperature range to approximately up to 300 °C.

In DTA, as in thermogravimetry, special methods of temperature measurement and thermocouple manufacture are also met. King and his co-workers [66] developed for DTA a thermocouple composed of thin layers of thermocouple metals, fixed on a quartz plate by condensation of metal vapour in vacuum. This is a microanalytical method, and the base plate on which the thermocouple metals are condensed, through a stencil in vacuo, has the dimensions $16.5 \times 13.7 \times 0.26$ mm. The differential thermocouple is composed of Ni and Au layers, and samples in aluminium foil vessels are placed at the junction. This arrangement allows work within a temperature interval from -125 to 500 °C and in view of the low weight of the thermocouple and the sample it gives good quantitative results. To increase the signal of the differential thermocouple, electronic amplifiers are often used. To avoid complications due to the simultaneous amplification of the noise, thermocouple systems composed of a large number of welded joints have been used [62].

The effect of the thermocouple itself may be summarized as follows.

(a) Effect of the position of the thermocouple (depth, volume immersed in the sample).

(b) Effect of the removal of heat by the thermocouple arms.

(c) Effect of the thermocouple on the sample (inertness of the thermocouple, influence on the process of the change).

It is necessary therefore to choose both the material and the dimensions of the thermocouple carefully (this also applies to the material of the crucible and the block), as well as the method of location of the sample. The importance of a symmetrical location of samples with respect to the heat source, as well as of a symmetrical location of the junction of the differential thermocouple with respect to the sample and the standard, has been shown in practice many times. It is recommended that the junctions are placed in the sample and the standard in exactly the same way, i.e. that their positions be either in the centre or at the same distance from it. An asymmetrical position produces a distortion of the curves, poor reproducibility, and makes a correct evaluation of the thermograms more difficult.

(d) The system of DTA curve recording. Garn [47] has discussed the question of whether to record ΔT against T, or whether to use a two-pen recorder to plot both T and ΔT independently as functions of time. He prefers the latter if a straight choice must be made, on the grounds that there is a check on the performance of the control equipment, and that the chart-speed can be varied at will so as to expand the record of rapid events and telescope that of periods when no reaction is

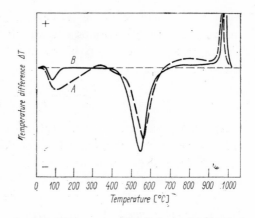

Fig. 3.12 DTA curves of kaolinite under identical experimental conditions. A — a photographic record with mirror galvanometer, B — registration by compensation recorder. Effect of dehydration at 100 °C is extended in the first case over the interval 30—300 °C and the main effect is shifted to higher temperature (Reprinted from [78] by permission of the copyright holder).

occurring. Berg [15a] has discussed some errors that may arise if ΔT is recorded against T. Sufficient precision and sensitivity of these measurements is ensured by the use of thermocouples such as those described in Section 2.4. Previously the temperature difference between the standard and the sample was most often recorded by using mirror galvanometers. More recently the ΔT signal is amplified electronically and recorded directly by eletronic compensation recorders. This second method of measurement is better, since some galvanometers have a slow response. The DTA curves for kaolin [78], shown in Fig. 3.12, may be used to compare the two methods of recording.

3.4.2. The method

(a) Nature of the sample. The problem of the geometry of the block and the cavities or crucibles for the location of samples is closely connected with the problem of the quantity and shape of the sample. The equation giving the peak area has already been given:

$$B = \frac{Q \cdot \varrho \cdot r^2}{4\lambda} \qquad (3.17)$$

Note the occurrence of the expression $\varrho r^2/\lambda$, which also occurs in the majority of the relationships derived for the characteristics of the DTA curve. Here ϱ is the density of the sample. For cylindrical symmetry this expression is defined by

$$\frac{\varrho r^2}{\lambda} = \frac{Mr^2}{V\lambda} = \frac{M}{\pi \cdot h\lambda} \qquad (3.21)$$

where M is the mass of the sample, V is its volume, h is the height of the cylinder and r its radius. The coefficient of thermal conductivity of the sample, λ, depends on its density, which in turn depends on its mass and dimensions. Thus, all the quantities are interdependent, as are changes in them, and the effect of the magnitude and dimensions of the sample should be considered from this point of view. For cylindrical samples a series of relationships has been deduced for the calculation of the peak area (B) when the value of the dimensions of the cavity (or the crucibles) and the quantity of the sample change [123]. Some of the practical consequences will be discussed to demonstrate the importance of the value of the density.

The area under the curve is proportional to the mass of the sample provided its density and thermal conductivity do not change. The relationship between these latter two is given, according to Jong [63], by the equation

$$\lambda = A + B\varrho \qquad (3.22)$$

where A and B are constants characteristic of the given system. The less the coefficient of thermal conductivity of the sample depends on its density, the more linear is the relationship between area, B, and the sample weight, M. Analogous considerations may be deduced for the amplitude of the temperature effects, i.e. for the relationship between its magnitude and the magnitude and dimensions of the sample.

From practical experience several conclusions may be deduced concerning the effect of the amount used and the dimensions of the sample. Commonly, sample amounts range from ten to several hundred milligrams.

In the early development of this method, the amount of sample used for some special applications (e.g. the analysis of metals and steels) was very large, e.g. up to 200 g. Increasing the weight of the sample, with the aim of obtaining larger values for the area of the effect, has its limits, however, and may lead to undesirable broadening of the curve and to superimposition of peaks. This means that the heating rate must be reduced. On the other hand, decreasing the sample weight leads to an increased ability to separate peaks and enables the heating rate to be increased. Decreasing the sample weight is limited, however, by the sensitivity of the apparatus. The lower the weight of the sample, the more pronounced will be the effect of the properties of the thermocouple on the distribution of temperature in the sample. With commonly used systems with a differential thermocouple arrangement it is possible to deal with samples weighing several tens of milligrams. However, there are special procedures (considered later) which deal with milligram quantities.

The amount of the sample may affect the position of the peak on the temperature axis and is closely connected with the place where the temperature is measured for recording the temperature axis, as was explained in the preceding section (p. 101). The position of the curve depends also on the type and the kinetics of the change. In the case of a first-order phase transition, when the temperature is measured in the sample the temperature of the transition coincides with the peak temperature (T_m). However, if the temperature is measured in the reference material the T_m value will depend on the sample mass. In chemical reactions, if the reaction order is equal to unity, the temperature of the peak is independent of the sample mass. But if the reaction order is less than unity, which is usually the case for the decomposition of solid substances, the peak is narrowed by an increase in the value of the start (T_0) and a decrease in the value of the peak temperature (T_m). This may mean that effects are partially superimposed.

Effects resulting from the packing of the sample may be quite substantial, though normally not enough attention is paid to them. The effects arise because of the relationship between the density, thermal conductivity and shape of the sample. Packing can substantially affect the heat transfer to the thermocouple junction, and the diffusion of gases through the sample. Close packing improves the heat transfer and prevents the diffusion of gas, while loose packing has the opposite effect. It is recommended that the sample and the reference material are packed in the same manner, otherwise deformation of the curve may occur as a result of a difference in thermal conductivity, although of course this does not depend exclusively on the degree of packing. In decomposition and redox reactions, there is a relationship between packing and the furnace atmosphere and whether a dynamic or a static atmosphere is used. Close packing, although it facilitates the heat transfer and thus increases the temperature effect, may, on the other hand, also affect the reaction itself, by affecting the diffusion processes. The method of packing will therefore vary from case to case, account being taken of the type of change observed and the reproducibility of the results.

(b) Effect of the furnace atmosphere. The furnace atmosphere has a substantial effect on reactions in which gases are liberated or in which the sample reacts with the components of the atmosphere. The effect of pressure on the DTA curve has already been mentioned in several of the preceding sections. Thermodynamically, a reaction giving a gaseous product (for example carbon dioxide), and which occurs in a gaseous medium with a given partial pressure of this gas, occurs only when the dissociation pressure of the decomposition reaction corresponds to the partial pressure of the gas or exceeds it. As the partial pressure increases, the dissociation temperature is shifted to higher values. This follows from the fundamental van't Hoff relationship:

$$\frac{d(\Delta G^0/T)}{dT} = -\frac{\Delta H^0}{T^2}$$

Now as

$$K_p = P_{CO_2} \tag{3.23}$$

and

$$\Delta G^0 = -RT \ln K_P = -RT \ln P_{CO_2}$$

i.e.

$$\frac{\Delta G^0}{T} = -R \ln P_{CO_2} \tag{3.24}$$

then

$$\frac{d \ln P_{CO_2}}{dT} = -\frac{\Delta H^0}{RT^2} \tag{3.25}$$

where K_p is the equilibrium constant of the reaction and ΔH^0 is the heat of reaction.

Every substance is stable only under given conditions of pressure, temperature, and atmospheric composition. According to the second law of thermodynamics, if any of these variables is changed, the system is also changed and will assume a new equilibrium state. By means of DTA all reactions may be followed which liberate or absorb heat. The change in energy during the reaction, referred to the unit of weight or volume of the given

system, may be the result of one or more reactions proceeding simultaneously. The reaction may be affected by changes in the variables, determined
in practice by experimental conditions (composition of the atmosphere,
pressure of its components, temperature, etc.). The possibility of applying
changes in these parameters depends on suitable apparatus.

Current practice in DTA usually does not peimit a precise definition
of the composition of the furnace atmosphere with respect to partial
pressures of its components. This is due to the fact that the composition
of the furnace atmosphere changes during the course of the experiment,
as a result of the reactions occurring and the streaming of gas through
the furnace. For this reason, methods were introduced which use a dynamic
atmosphere, where the gas stream is led through the furnace or directly
through the sample, for example via the porous bottom of the crucible
(see Fig. 5.1). A dynamic atmosphere is established under controlled
conditions, e.g. Stone's apparatus [146–149], which permits control of all
three variable factors. He uses a stream of carrier gas passing directly
through the analysed sample, in contrast to earlier methods [131] which
have the dynamic atmosphere only outside the sample. It will be shown
in Section 3.7 that this arrangement provides other possibilities for
quantitative evaluation of results. Another method by which the reaction
conditions may be affected is working with reduced pressure. The adjustment of experimental conditions in this way depends on the aim of the
experiment, e.g. in the case of kinetics, there is the problem of the nature of
the controlling process. Only some general aspects and experimental
possibilities will be presented here. Garn [47] gives a more detailed
treatment.

When a dynamic atmosphere is used, different aspects of the given
system can be studied by suitable arrangement of experimental conditions,
e.g. when an inert gas is used only the effect of its pressure can be
investigated. On the other hand, if a gas is used which plays a direct
part in the reaction under investigation, the relationships following from
the basic van't Hoff and Clausius–Clapeyron equations may be studied.
The effect of static and dynamic atmospheres on reactions occurring during
the thermal treatment of ores has been extensively studied [132, 149].
It is seen that the value T_m of the endothermic peak is shifted to lower
temperatures when a dynamic atmosphere is used, but that the shape of
the curve remains unchanged. When different compositions of furnace
atmosphere are used, both for static and dynamic conditions, the most
important factor is the type of gas, e.g. the use of gases such as nitrogen,
argon, or carbon dioxide prevents oxidation. The curves of siderite and
$MnCO_3$ serve as a typical example in which the exothermic effect of FeO

and MnO oxidation (following immediately after the decomposition of carbonate) may disappear completely in a neutral CO_2 atmosphere (Fig. 3.13). In contrast to this, when carbonates which yield decomposition products which do not undergo further oxidation, e.g. dolomite or calcite, are heated in a CO_2 atmosphere, the increase of the T_m value is much more distinct, and the deviation from the zero line is much sharper. This is

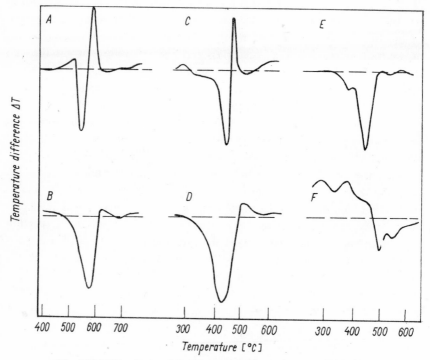

Fig. 3.13 DTA curves of $FeCO_3$ and $MnCO_3$ (in different gases).
A — $FeCO_3$ in air: B — $FeCO_3$ in CO_2: C — $MnCO_3$ in air: D — $MnCO_3$ in CO_2: E — $MnCO_3$ in N_2: F — $MnCO_3$ in O_2 (Own measurements)

due to the higher temperature required to attain the equilibrium pressure. In an oxygen or air atmosphere oxidation reactions occur as shown in Fig. 3.14. In addition to the technique of the dynamic atmosphere in DTA, which was developed mainly by Stone and later by Lodding and Hammell [86], varying the pressure of the furnace atmosphere [149] is also used. Garn [49] used this technique for the study of dehydration of barium chloride in the medium of its own atmosphere at a pressure of up to 8 atmospheres. With increasing pressure, the effects of dehydration usually shift to higher temperatures. The importance of the methods using a controlled atmosphere is predominantly in the determination of suitable conditions for reactions. The use of various static atmospheres in reactions which liberate the particular gas or react with it, gives valuable information in the case of reversible and also some irreversible reactions. The use of dynamic atmospheres permits easier identification of the reaction type and of the gas liberated, as well as a quantitative evaluation — as will be shown in Section 3.7.

In connection with the problem of furnace atmosphere it should be pointed out that a rather frequently used technique consists in sealing the vessels containing the samples (see Figs. 5.1 and 5.2). The main effect of this procedure is to prevent diffusion and escape of the gaseous reaction products as well as preventing access of the ambient atmosphere to the sample. Similar effects can also be produced by covering the sample with a layer of inert material. However, in current practice, DTA is often carried out in an uncont rolled atmosphere, i.e. directly in the

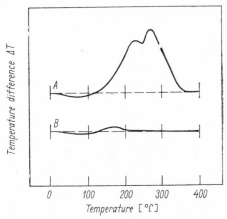

Fig. 3.14 Effect of atmosphere on the shape of DTA curves of Ni catalyst.
(62.5% Ni; 8.1% Al) Ni-powder as reference.
A — in air: B — in H_2 atmosphere
(author's measurements)

laboratory atmosphere. This atmosphere will contain varying quantities of moisture and carbon dioxide, and may influence the temperature of decomposition of a carbonate by as much as several tens of degrees [3].

Thus it is seen that it is important to consider the effect of all three variables, viz. temperature, pressure, and the composition of the gas phase, on the reactions, and thus also on the shape of the DTA curve. In practice two variables are usually kept constant and the third is changed. According to Stone [149] it is possible to make use of five basic procedures for combining the variable parameters (see Table 3.1). No. 2 is applied when two dynamic atmospheres are formed in the course of the process. No. 3 is difficult to carry out and therefore No. 1 is usually made use of at various pressures. With No. 4, the reaction which causes the temperature effect is started as a result of a change in pressure, but not in temperature. The determination of moisture in powdered materials is an example; this is carried out by evacuating the system at a constant rate at room temperature. The deviation on the DTA curve starts when the pressure reaches the value of the water vapour pressure at room temperature. The decomposition of carbonates has been investigated in a similar manner. No. 5 utilizes an abrupt change in atmosphere, causing a sudden change in equilibrium conditions. The substance is transferred suddenly from a medium in which it is stable to one in which it is unstable, and therefore the reaction occurs rapidly. This method is not widespread, but the reactions studied may be accurately defined; they concern mainly redox

Table 3.1

Basic Procedures of DTA

Procedure No.	Temperature	Pressure	Gas composition
1.	Increasing at constant rate	Constant	Constant
2.	Increasing at constant rate	Constant	Variable for a given temperature
3.	Increasing at constant rate	Variable for a given temperature	Constant
4.	Constant	Variable	Constant
5.	Constant	Periodic or constant	Changing periodically

reactions, studies of catalysts and studies of surface activity. The effect of furnace atmosphere may be summarized as follows.

(a) The pressure and composition of the furnace atmosphere may affect the equilibrium temperature both of chemical reactions and of physical changes. In the case of physical transformations which occur without involving the gas phase, the T_m value is only slightly dependent on the atmosphere pressure. In chemical reactions or in the case of physical changes in which the gas phase takes part (e.g. decompositions, sublimation, dehydration) the value of the equilibrium temperature is strongly pressure-dependent.

(b) The composition of the furnace atmosphere may lead to a reaction between the analysed sample and its decomposition products, solid and gaseous. Oxidation and reduction are among the most common chemical reactions of this type. At different pressures the equilibrium temperature of reversible reactions is shifted. The diffusion of the gaseous products of decomposition as well as the diffusion of the atmosphere to the sample may be affected by experimental conditions (e.g. weight of the sample, geometry of the sample, dilution, and rate of heating).

(*c*) The composition of the furnace atmosphere and the partial pressure of the gaseous components may affect the mechanism of the reaction in spite of the fact that they do not play a direct part in the reaction, e.g. at lower pressure metastable phases may be formed.

(c) Effect of the rate of heating. The rate of heating is obviously of importance, since the rate of reaction will depend on the temperature of the sample, and if the temperature rise from external sources is considerable during the lifetime of the reaction, the DTA curve will differ from that obtained at a different rate of heating. The situation is further complicated by the rate of diffusion of evolved gases — the higher the heating rate, the less diffusion can occur in the time needed to traverse the temperature span of the thermal effect studied, and the greater the effect of the gaseous atmosphere on the equilibrium of the reaction. At low heating rates a true equilibrium will be more nearly approached, and extrapolation to zero rate of heating has been used in certain applications [10, 154]. Figures 3.5 and 3.6 (pp. 94 and 95) illustrate the effect of heating rate on the shape of DTA curves, and the effect on evaluation of the curves is discussed on pp. 93 ff.

The rate of heating should be as linear as possible, to avoid unwanted side-effects from its variation, and the apparatus should be arranged to give as efficient heat transfer as possible, to avoid temperature lag between sample and reference material. This problem of heat transfer is again closely connected with the nature and packing of the sample, the construction of the block, etc. (see pp. 78, 108).

3.4.3. Sample and reference material

Effect of physical and chemical properties of the sample

An important but little studied factor affecting the DTA curve is the particle size, and the results given in the literature are often contradictory. This is caused by the fact that the problem of particle size is not a simple one; it should be considered not only from the purely physical point of view, but also from the point of view of other conditions such as the surface reactivity, crystal structure, history of the formation of the particles (crushing and grinding), etc. Speil and co-workers [143] found that during the dehydration of kaolin an increase in particle size from 0.1 to 20 µm increased the area under the curve by a factor of almost three. However, the results of other authors [54, 114] do not agree with this. Discrepancies are also met in the evaluation of the effect of the particle size on the T_m value.

A smaller particle tends to react more readily than a larger one, the reason being that it more often has an imperfect surface. With smaller crystalline particles, the number of active sites, such as edges and corners, is greater. The magnitude of the active surface, determining the reactivity of the substance, increases with decreasing particle size. Smaller particles may be obtained, in principle, in two ways; either by hand-picking or sieving the original material or by decreasing the size by physical means. In the second case the product obtained displays a greater reactivity. This can be explained by the fact that in the case of the unground material a decrease in the irregularities occurs. The methods of physically reducing the particle size, such as wet or dry grinding, or crushing, will naturally have an appreciable effect on the resulting reactivity of the product. The effect of the size of the particles on the DTA curve may thus be related directly to the method of obtaining that size of particle. A gradual diminution in size may have an appreciable effect, mainly in the case of dehydration reactions in which it was observed that with progressive grinding (and hence smaller particle size) the deviation due to dehydration was decreased and was shifted to lower temperatures. After wet grinding of the sample the DTA curves are often simpler and better shaped. This is also true of the dehydration curves obtained with a material ground in a medium of increased water vapour pressure. Wet grinding leads to substantially smaller surface irregularities than dry grinding, and the latter may lead to changes on the surface of the particles (for example the formation of surface films) that may affect changes accompanied by a small heat effect [38], e.g. it was found that grinding increased the defect structure of hydrargilite [152]. Bando and co-workers [6] found that during the transformation of γ-Fe_2O_3 to α-Fe_2O_3, a decrease in grain size altered the temperature of the change. The majority of reactions in the solid phase are controlled by diffusion and it may be expected that this mechanism will be affected by the particle size. Another factor is the effect of crystallinity, i.e. the degree of perfection of the crystal structure. The clarity and sharpness of the observed effects increases with the degree of crystallinity. Particle size may give rise to quite new effects. Thus when DTA is applied to glasses, when finely ground samples are employed an exothermic effect appears on the DTA curve in the region of 700–800 °C which corresponds to the sintering of the particles in the sample. When a compact sample is heated the effect disappears.

Effect of the physical and chemical properties of the reference material

In DTA, when no reaction is occurring in the sample, ΔT should be minimum and the baseline should have minimum slope. This is the case only when the thermal conductivity of the reference material corresponds to the thermal conductivity of the analysed sample over the whole temperature range investigated. Thermal conductivity – and hence, also, the fulfilment of this requirement concerning the baseline – depends on the basic physical properties of these materials, such as specific heat, density, particle size, and packing. Hence, it is evident that the deviation of the curve from the baseline is another characteristic describing the DTA curve; this is shown, under ideal conditions, by the curve in Fig. 3.3. In practice, the DTA curve usually approaches the zero line satisfactorily and sometimes even merges with it, which is the aim of most experimenters. In principle, this coincidence may have two causes. Either the sensitivity of the apparatus is so low that deviation from the zero line cannot be recorded, or the pair of samples compared was chosen so that their thermal diffusivities were equal ($a_{sm} = a_{st}$). In the latter case the following relationship holds [following from Eq. (3.18)].

$$\frac{M_1 c_1}{M_2 c_2} = \frac{V_1 \lambda_1}{V_2 \lambda_2} \qquad (3.26)$$

where M_1 and M_2 are the weights, V_1 and V_2 the volumes, c_1 and c_2 the thermal capacities and λ_1 and λ_2 the coefficients of thermal conductivity of the sample and reference material respectively.

From this equation it follows that the parameter which will have the greatest effect on the deviation of the curve (ΔT) will be the coefficient of thermal conductivity when the volumes and the weights of both sample and reference material agree as closely as possible. It can be deduced that the DTA curve will merge with the zero line if

$$\frac{\lambda_2}{\lambda_1} = \frac{M_2 c_2}{M_1 c_1} \qquad \text{and} \qquad \Delta T = 0 \qquad (3.27)$$

be above the zero line if

$$\frac{\lambda_2}{\lambda_1} < \frac{M_2 c_2}{M_1 c_1} \qquad \text{and} \qquad \Delta T > 0 \qquad (3.28)$$

and below the zero line if

$$\frac{\lambda_2}{\lambda_1} > \frac{M_2 c_2}{M_1 c_1} \qquad \text{and} \qquad \Delta T < 0 \qquad (3.29)$$

Obtaining total coincidence between the thermal conductivities of sample and reference material is practically impossible, as some of the parameters which determine the thermal conductivities change during the course of the investigated transformation and the observed temperature range. However, a small slope in the zero line, especially in the case of modern sensitive apparatus, is permissible even in the case of quantitative methods, as will be shown in the methods of graphical evaluation of the curves. For these reasons, large deviations of the curve from the zero line (drift) are met, even in the case of well-shaped temperature effects. Very often this drift is eliminated by decreasing the sensitivity of the apparatus, but this serves little purpose. A more useful procedure is to find a suitable reference material, to fulfil the condition of Eq. (3.27).

Calcined alumina, MgO, and fused quartz are most commonly used as reference materials. In special cases various other materials are also used which are chosen quite empirically with respect to the sample analysed, e.g. powdered nickel was used in the analysis of the electrodes of fuel cells [20]. However, the basic requirement for the deviation of the curve from zero line is its reproducibility. It should be kept in mind that the ΔT value in the steady state is directly proportional to the rate of heating [3]:

$$\Delta T = \frac{dT}{dt} \cdot r^2 \left(\frac{1}{a_1} - \frac{1}{a_2} \right) \Big/ 4 \qquad (3.30)$$

where dT/dt is the rate of heating, r is the radius of the cavity, a_1 and a_2 are the coefficients of thermal diffusivity of both materials. This means that increasing the heating rate will increase the drift of the curve from the zero line. In quantitative analysis and theoretical investigations, a careful choice of the reference material is recommended, as well as of the material which is used for possible dilution of the analysed sample. Thus, it can be said that the faster the heating and the greater the difference in weight, radius or thermal properties between the sample and the standard, the greater will be the baseline drift.

Effect of dilution of the sample with an inert material

As was indicated in the preceding section, the reference material may often be used for the dilution of the analysed sample—for the following purposes.

1. To prepare samples of various concentrations of the active substance in quantitative analysis.
2. To prevent sintering of the sample.
3. To decrease the recorded temperature effect.
4. To affect the contact between the sample and the ambient atmosphere.

5. To reduce baseline drift.

6. To adjust the thermal conductivity of the sample.

7. For special microanalytical procedures.

During the thermal treatment of the sample its heat capacity usually changes as a result of physical or chemical changes. The use of diluents may enable the heat capacity to be kept at a relatively constant value, if the heat capacity of the sample is small compared with the heat capacity of the diluent. Using a diluent of high heat capacity decreases the peak height while an increased thermal conductivity of the diluent increases it. Dilution of the sample with an inert material should not be done without taking into account all possible complications which it could produce. This is especially true of those reactions which are affected by the composition of the furnace atmosphere, mainly by the access of oxygen. The curves of the decomposition of siderite mixed with Al_2O_3 (see Fig. 3.15) [133] are a typical example. A similar set of curves for the decomposition of $MnCO_3$ in the presence of alumina were observed by the author. At a low diluent concentration the atmospheric oxygen diffuses to the sample at a rate sufficient to create an energy-rich oxidative effect. Increasing the content of diluent results in compensation of effects, until the decomposition effect disappears completely; there will be an optimum value for the amount of diluent added. When a dynamic atmosphere is used, ensuring constant composition of the atmosphere in the vicinity of each particle, the effects of the diluent are eliminated. The dilution of the sample with an inert material may also lead to changes in the T value. Dean [37] observed a shift of T_m of the endothermic peak

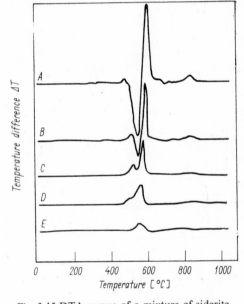

Fig. 3.15 DTA curves of a mixture of siderite with alumina. The magnitude of the effect changes with varying access of oxygen to the sample.

A — 40% $FeCO_3$ and 60% Al_2O_3.
B — 35% $FeCO_3$ and 65% Al_2O_3.
C — 33% $FeCO_3$ and 67% Al_2O_3.
D — 30% $FeCO_3$ and 70% Al_2O_3.
E — 26% $FeCO_3$ and 74% Al_2O_3.

(Reprinted from [78] by permission of the copyright holder)

of kaolin to higher temperatures when the amount of diluent Al_2O_3 was increased. In the case of other diluents (e.g. Ni, Fe) [11, 20] the main effect is on the thermal conductivity of the material. An unsuspected difficulty may be adsorption of evolved gases by the diluent, followed by desorption at a higher temperature.

3.5. METHOD OF TEMPERATURE MEASUREMENT AND CALIBRATION OF TEMPERATURE AXIS

Compared with TG, DTA usually presents a greater danger of contamination of the thermocouple, as this is frequently placed in direct contact with the sample. Therefore, more frequent calibration is necessary, defining the temperature difference between the sample and the site from which the temperature curve is usually taken usually not the sample, but the block or the reference material. Just as the temperature curve in TG may be calibrated by using internal standards, this method is often used in DTA, by adding a small amount of a substance with a well-defined temperature effect to the analysed sample or the standard. This calibration method was perfected by Barshad [13]. Although somewhat tedious, this procedure affords the possibility of an accurate determination of the temperature of the observed effects when the linearity of heating cannot be sufficiently assured. The method also permits a simultaneous evaluation of the sensitivity of the apparatus, which is important in quantitative analysis. A large num-

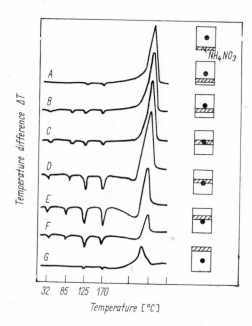

Fig. 3.16 Effect of the location of the active substance with respect to the position of the thermocouple junction. Curves of mixtures of NH_4NO_3 (15 mg) and Al_2O_3 (x mg)

$A - x = 500$ mg; $B - x = 75$ mg;
$C - x = 150$ mg; $D - x = 225$ mg;
$E - x = 275$ mg; $F - x = 325$ mg;
$G - x = 500$ mg;

(Reprinted from [13] by permission of the copyright holder)

ber of these internal standards suitably distributed in both sample and reference may be used, so that their effects may cover as large a range of temperature as possible. Barshad demonstrated that by the application of this calibration method an appreciable influence on a distinct peak is exerted only by a sample situated in the immediate vicinity of the thermocouple junction, i.e. when the active substance is in contact with the thermocouple. This is shown by the curves of the Al_2O_3 and NH_4NO_3 mixture shown in Fig. 3.16. When the internal standard is applied in close proximity to the thermocouple a very small amount of the sample (2−20 mg) is sufficient. This is usually used to advantage in microanalytical methods.

3.6. PRETREATMENT OF THE SAMPLE

Pretreatment of the sample for analysis is a factor which may appreciably affect the curve. This factor should be considered from two points of view, viz. chemical preparation proper, carried out before the experiment, and location of the sample in the crucible. The chemical preparation of the sample before the experiment is aimed at simplifying the curve and eliminating some disturbing effects caused by the presence of impurities. For example, organic substances or carbonates are the most common impurities in minerals; to remove these impurities some workers use extraction with acid or hydrogen peroxide [48]. Another factor in the standardization of mineral samples is the humidity, which can be regulated by keeping the samples over a saturated solution of magnesium nitrate. The choice of the method of standardization or pretreatment of the sample depends mainly on the purpose of the thermal analysis. If the analysis is to determine the basic properties of pure materials, it is convenient to pretreat the sample by procedures which decrease the content of impurities. However, if the analysis is to determine the composition of the sample or to estimate the impurities present, then obviously pretreatment procedures are not permissible. The grinding of material and its effect on the DTA behaviour of the substance was discussed on p. 116.

The second factor in the pretreatment of the analytical sample, its position in the crucible, is discussed on p. 99. In common practice various methods of packing the sample are used. The degree of packing is also very important from the point of view of the reproducibility of the curves, mainly in quantitative analysis. Packing, together with the thermal conductivity and the heat capacity of the sample, is temperaturedependent and may change during the experiment; this is usually made evident from a baseline drift.

3.7. QUANTITATIVE DIFFERENTIAL THERMAL ANALYSIS

The extensive development of DTA in recent years has brought about a tendency to express DTA curves in mathematical terms and to make use of them for quantitative determination of the heat of reaction or of other changes occurring. The quantitative aspect of the method is that a measure of the heat of reaction and hence also of the amount of the active substance in the sample can be made from a consideration of some characteristics of the curve, mainly of the area under the curve, expressed by its integral. However, it should be noted that the results of quantitative evaluation depend on the conditions and factors which have been discussed in the preceding sections, and that the majority of the mathematical expressions for the DTA curves are more or less approximate and often based on assumptions which are not always completely valid. This does not mean that quantitative applications of DTA are meaningless, but it shows that they have a limited validity and that all factors should be considered carefully from case to case.

As was shown in the preceding sections, the DTA curves may be defined by a series of mathematically expressible characteristics (e.g. the deviation of the curve from the baseline, temperature of the start and the peak of the thermal effect, area and amplitude of the peak) of which some may be used for quantitative evaluation of the curves. The area under the curve and the peak amplitude of the deviation from the baseline are the most important of these, although some other characteristics may also be evaluated quantitatively. The first step in the evaluation is the determination of the relationship between the material's properties and the curve characteristics mentioned. This relationship may be determined experimentally or analytically. The second method requires the solution of differential equations. This is facilitated by the introduction of simplifying assumptions which enable them to be solved as equations of thermal conductivity under quasi-stationary conditions. The following five simplifying assumptions are made [123].

(a) Thermal physical properties of the sample (a, c, λ) do not depend on temperature, and after reaching the quasi-stationary state the temperature curve becomes linear.

(b) Before and after the transformation the sample has the same thermal physical properties, so that the beginning and the end of the temperature effect are at the same level.

(c) At the moment of the change, the concentration gradients of the components of the change are so small that they may be neglected; they do not affect the temperature regions in the sample by mass transfer.

(d) The change follows the basic kinetic equation

$$\frac{dx}{dt} = f(x) \cdot e^{-E/RT} \tag{3.31}$$

where x is one of the quantities expressing the degree of the change.

(e) The change occurs under quasi-stationary conditions.

It was determined experimentally that the heat of reaction as well as the mass of the sample or of its thermally active part are directly proportional to the peak area. From Eq. (3.6), which was deduced for the peak area, it is evident that the area, B, is directly proportional to ΔH or the active mass of the sample M_a, if the other parameters, i.e. the coefficient of thermal conductivity and the geometrical arrangement, remain unchanged. The majority of the methods used for the determination of the active mass and the heat of reaction from the peak area are comparative and empirical methods. A selected characteristic of the investigated sample (e.g. the value of B) is compared with the same characteristic of the standard substance. Obviously basic constants (λ, a, ϱ) and geometrical arrangements of both substances should be approximately equal, which is achieved by various experimental procedures (choice of suitable reference substances, use of equivalent crucibles, mixing of the sample and reference substance with a thermally inert material in equal parts, etc.).

For quantitative evaluation of the DTA curves a series of mathematical relationships has been derived. Some of these were discussed in Section 3.3, as for example those expressed by Eqs. (3.6), (3.8), (3.10), and (3.11). In quantitative analysis the measure of the amount of the active substance or of the heat of reaction is given by the area under the curve (B), obtained by integration of the DTA curve between the beginning of the effect (T_2) and its end (T_3), as shown in Fig. 3.3. The area under the curve is the area limited by the baseline and the curve of the temperature difference recorded as a function of time, according to the equation

$$B = \int_{t_2}^{t_3} \Delta T \, dt = \Delta Q \cdot A \tag{3.32}$$

where A is the proportionality constant determined experimentally, or calculated on the basis of apparatus parameters [25, 63] comprising the calibration factor and the density of the sample, ΔT is the temperature difference between the sample and the reference, ΔQ is the heat of the reaction (in cal/g). This equation applies when the heating rate is relatively slow and constant, permitting attainment of a quasi-stationary state.

The proportionality constant, A, depends on the apparatus, temperature, and the sensitivity of the method. For quantitative evaluation it is usually determined experimentally, i.e. either by calibration with substances of known heats of reaction, or by internal heating for calibration. Substances suitable for this calibration are given in Section 4.2.3. It should be noted that these calibration curves display an increasing value of A with increasing temperature, as shown in Fig. 4.1. This increase of the calibration curve is related to the increasing proportion of heat transfer by radiation. For a relatively narrow temperature interval the value of A may be considered constant, and its average value may be substituted into Eq. (3.32). Calibration may also be done electrically by using a calibration heater which is placed inside the cavity with the inert substance; when a normal recording of the DTA curve is made, a known amount of heat is transferred to this sample at chosen temperatures. The ratio of the heat transferred in this manner to the area under the DTA curve determines the value of factor A at given temperatures. The calibration of the experimental arrangement, which is necessary for quantitative evaluation of temperature effects, may be carried out simultaneously during every quantitative determination by using a suitable substance of known heat of reaction instead of an inert reference material, which means that in addition to the effect of the analysed substance, an opposing effect of the calibration substance is also observed; the heat of change of the analysed substance can be calculated from the ratio of the areas of both effects, multiplied by the heat of change of the calibration substance.

Some workers have attempted to verify experimentally that the area under the curve corresponds to the heat of reaction [15, 63, 70, 138]. The checking of these relationships is possible if the liberated reaction heat is measured and if an equation is available which expresses this process. One of the first theories was proposed by Speil [142], according to whom the area of the peak is given by the equation

$$B = \int_{t_2}^{t_3} \Delta T \, dt = q \cdot \frac{M}{g\lambda} \tag{3.33}$$

where q is the heat of reaction (cal/g), M is the total weight of the sample (g), g is the geometrical factor on which the temperature gradient in the sample depends, and λ is the coefficient of thermal conductivity of the sample.

If the value of $M/g\lambda$ is known, the value of q may be calculated. This expression is in fact identical with the proportionality constant from Eq. (3.32), which may be expressed by the equation

$$A = \frac{\text{area of the peak}}{q} = \frac{M}{g\lambda} \qquad (3.34)$$

This calibration factor was accurately determined by various authors. Kronig and Snoodijk [73] based their determination on the theory of heat conduction in a sample of cylindrical symmetry and found that

$$A = \frac{\varrho \cdot a^2}{4\lambda} = \frac{M}{4\pi \cdot h \cdot \lambda} \qquad (3.35)$$

For a spherical sample the following equation was deduced:

$$A = \frac{\varrho \cdot a^2}{6\lambda} = \frac{M}{8\pi \cdot a \cdot \lambda} \qquad (3.36)$$

where ϱ is the sample density (g/cm^3), a is the sample radius (cm), and h is the height of the sample (cm).

These relationships describe the temperature difference between the centres of the analysed and the reference samples assuming that the differential thermocouple is very thin. Boersma [25] considered the effect of heat losses caused by removal of heat through the thermocouple and arrived at the following expression for spherical symmetry

$$B = \frac{qa^2}{6\lambda} \cdot \frac{\alpha}{1 + \Lambda/\lambda} \qquad (3.37)$$

where

$$\alpha = 1 - (r_0^2/a^2) [3 - 2(r_0/a)]$$

and

$$\Lambda = \lambda_p(r_0/l) (A/4r_0^2) [(1 - r_0/a)]$$

where r_0 is the radius of the thermocouple junction, a the radius of the cavity filled with sample, λ_p the coefficient of thermal conductivity of the thermocouple wires, l the length of leads in which the temperature gradient exists, A the cross-sectional area of the leads, λ the coefficient of thermal conductivity of sample material, q the heat of transformation per unit volume, and for cylindrical symmetry he obtained the exprassion

$$B = \frac{qa^2}{4\lambda} \cdot \frac{1 - \dfrac{r_0^2}{a^2}\left(1 + 2\ln\dfrac{a}{r_0}\right)}{1 + \dfrac{A}{l}\dfrac{\lambda_p}{\lambda}\dfrac{\ln(a/r_0)}{2\pi h}} \qquad (3.38)$$

where h is the height of the cylinder.

These relationships do not take into account the rate of heating as a factor affecting calibration.

Melling and co-workers [171] studied the effects of samples and the block on the heat transfer and on a mathematical model for the expression of heat transfer. These workers took as their model a massive block of high thermal conductivity with cylindrical holes and considered the decrease in the area under the curve as a result of heat losses caused by the thermocouple (according to Boersma) and they found that the decrease was larger in samples of spherical symmetry (50%) than in those of cylindrical symmetry (15%). Heat losses caused by the thermocouple decrease the peak area and the T_m values. Heat losses may be reduced by using thin thermocouple wires.

The relationship between the area under the curve and the amount of the substance analysed may be summarized by the following simple relationships.

For spherical symmetry

$$B = \frac{\Delta Q a^2}{6\lambda} \tag{3.39}$$

For cylindrical symmetry

$$B = \frac{\Delta Q a^2}{4\lambda} \tag{3.40}$$

For a sample of a general shape the relationship will be

$$B = \frac{\Delta H \cdot M}{\lambda \cdot k} \tag{3.41}$$

where ΔQ is the amount of heat liberated or absorbed by unit volume of sample, ΔH the amount of heat liberated or absorbed by unit weight of the active part of the sample, a is the radius of the spherical or cylindrical vessel, λ is the coefficient of thermal conductivity of the sample, M is the mass of the active part of the sample, and k is a constant dependent on the geometrical arrangement of the sample.

From these relationships it follows that the area under the curve is inversely proportional to thermal conductivity (and thus directly proportional to density) and directly proportional to the heat of reaction. The thermal conductivity of the sample is dependent on its packing and particle size. Thus two reactions with the same heat of reaction may give rise to different DTA effects, depending on the values of the coefficients of thermal conductivity. The relationships which show direct proportionality of the peak area to the heat of reaction and the mass of the sample are the basis of practical quantitative evaluations. However, it is important to realize that the quality of the results will depend on the degree of the control of the parameters affecting the curve, both those which may be included in the value of the constant calibration factor, and those which affect the result directly. One of these factors will be the temperature difference, ΔT, as a function of temperature which will be complex under conditions when heat transfer is by radiation. Thus, the calibration of the system for the whole temperature range used is of great importance.

A further possible quantitative use of DTA is the determination of specific heats. According to Adam and Müller [177] this can be done with a special calorimetric device due to Boersma (see Section 3.3 and Fig. 5.1 in Chapter 5), or according to Schwiete and Ziegler (see Section 3.9.4) by keeping one of the vessels of known heat capacity, c_n, empty. The unknown value of the specific heat of the investigated substance, c_{sm}, is calculated assuming a constant heating rate and neglecting the temperature gradient inside the sample, using the equation

$$\frac{c_{sm}}{K} = \frac{\Delta T}{\Phi} + \left[\frac{\Delta T}{\Phi} + \frac{c_n}{K}\right] \frac{d\Delta T/dt}{[\Phi - (d\Delta T/dt)]} \qquad (3.42)$$

where c_{sm} is the specific heat of the investigated sample, c_n is the specific heat of the vessel, Φ is the rate of heating, dT/dt, and K is a proportionality factor.

In temperature ranges where c_{sm} changes by only a small amount, it follows that $d\Delta T/dt \ll \Phi$, and therefore the simplified equation

$$\frac{c_{sm}}{K} = -\frac{\Delta T}{\Phi} \qquad (3.43)$$

can be used. In the region where the changes take place and when the c_{sm} value changes appreciably, Eq. (3.42) should be used.

Thus, it may be said that the DTA method may be used for the measurement of the enthalpy change (ΔH) and that theoretical relationships have been deduced expressing a relationship between the area under the curve and the heat of reaction. Several workers have made use of this method for the determination of heats of reaction, dehydration, melting, and transformation [41, 142, 143]. Usually the method of direct calibration of the apparatus is used with substances with known values for their heats of reaction. The relationship between the heat liberated or absorbed by the sample and the area below the curve will depend on the apparatus, which affects the heat flow inside the samples by virtue of the basic geometric arrangement. Most information about this relationship is given by the solution of an equation for the three-dimensional heat flow inside a homogeneous medium, modified for the limiting conditions determined by the given apparatus. Typical conditions for the solution of this equation are: (a) the thermal conductivity of the sample holder is much larger than that of the sample material, (b) the thermocouple is situated in the centre of thef sample, (c) the thermal conductivity of the sample and of the standard are approximately equal, (d) the liberation of heat in the sample is homogeneous with respect to the vertical axis. It is evident that some of these conditions are connected directly with the apparatus used; e.g. conditions (a) and (d) can be obtained by using a metallic block with holes of cylindrical symmetry. Other limiting conditions will apply, of course, for the solution of the equation when a ceramic block and different cavity symmetry are used. The condition (c) is usually obtained by diluting the sample with some inert material. The supposition that the liberation of heat should be uniform with respect to the vertical axis of the sample is fulfilled if the sample is symmetrical with respect to this axis and if it displays a uniform thermal conductivity over the whole volume. In this case the temperature gradient through the vertical axis is small, and may be further decreased by insulation of the upper and lower part of the sample by layers of Al_2O_3 (approximately 1.5 mm thick). The calibration by means of standard substances is usually carried out with substances suitable for the required temperature range. The peak area is evaluated in various ways, e.g. by cutting out and weighing. (For a more detailed discussion see Section 3.7.2.)

In addition to this method of measuring ΔH, another method is also possible, using the Clausius–Clapeyron equation

$$\frac{dP}{dT} = \frac{\Delta H}{T_p . \Delta V} \tag{3.44}$$

where ΔH is the latent molar enthalpy of the reversible change, T_p is the temperature of the change, ΔV is the change in molar volume during the change, and dP/dT is the change of the equilibrium pressure with temperature.

Under suitable experimental conditions the dependence of the T_p values on pressure can be determined during DTA. For a change in crystal modification in the solid state, only the T_p value at normal pressure is used. In reactions in which the gas phase takes part (for example a reversible chemical reaction or a decomposition), the volume, V, of the liberated gas phase is substituted for ΔV. At sufficiently elevated temperatures and low pressures, V can be expressed by the equation of state $PV = RT$, and the following expression is obtained:

$$\frac{dP}{dT} = \frac{\Delta H}{RT^2} \tag{3.45}$$

After integration, Eq. (3.46) is obtained for the calculation of the enthalpy:

$$\ln P = \Delta H/RT + \text{constant} \tag{3.46}$$

This method, used for the first time by Stone [148] in 1954, requires a special arrangement of the DTA apparatus, permitting control of the atmosphere and the presence of gas surrounding the sample. For this equation to hold, the following conditions apply: (a) the gas phase is an ideal gas, (b) the volume of the gas phase is much larger than that of the solid phase, (c) the system is closed, (d) the phase change is reversible, (e) there is only a pressure-volume change. Thus the gas phase surrounding the sample should be carefully controlled and the temperature of the beginning of the reaction and the reaction product must be known. This equation may be applied to a well-defined reaction liberating a gas of known composition, which may be introduced from an external source (dynamic atmosphere) under strictly controlled conditions. In practice, $\ln P$ is plotted against $1/T$, giving a straight line, with slope $\Delta H/R$, from which the enthalpy change, ΔH, is calculated. The use of this relationship is limited to reversible phase changes, which follows from the assumption that the chemical potentials of the gas phase and the solid phase are equal. Ellis and co-workers [40] applied both methods in the determination of the heat of decomposition of magnesium carbonate, with good results. A detailed review of the papers in which the Clausius – Clapeyron equation was used for the evaluation of enthalpy changes by DTA is given by Schultz [145].

3.7.1. Fundamental factors affecting quantitative evaluation of the curve

As thermal insulation between sample and reference is not necessary, the ΔT value decreases and eventually the temperatures become equal. This equalization of temperatures, which is time-dependent, depends also on the properties of the block (such as its thermal conductivity). With a slow heating rate the possibility of equalization is higher, and hence the magnitude of the temperature effect is smaller than on rapid heating. Therefore, the heating rate is of appreciable importance in quantitative application of the DTA curve. However, opinions as to the nature of the heating rate effect on the area under the curve differ [47, pp. 60, 137, 171]. If the area under the curve is plotted against the sample temperature, it is found that the area is proportional to the rate of heating if this does not change during the reaction.

Another factor important in quantitative analysis is the amount of sample. A temperature gradient exists inside the analysed sample, causing the heat of the reaction not to be absorbed or liberated at the same rate all over the sample, and thus the area under the curve is not necessarily always a linear function of the heat of reaction. This condition may be fulfilled in the case of sufficiently small samples, but in this case accurate temperature measurement becomes a problem. It has been demonstrated that the temperature of the peak, T_m, increases with increasing sample radius, density or specific heat and decreasing thermal conductivity. In connection with this, micromethods have recently become important.

Quantitative measurements of reversible transformations in which no liberation of gas takes place (e.g. the transformation of SiO_2) are usually not difficult, because the porosity of the material, rate of heating, etc., have no effect, and the transformation temperature is usually well-defined. However, the situation is different in the case of chemical reactions, such as decomposition reactions, which are accompanied by the liberation of gaseous products and thus also by changes of the atmospheric composition and with shifts in the temperature effects dependent on the heating rate, because the temperature of the reaction depends on the partial pressure of every gaseous product—as explained earlier. Therefore, the area of the peak corresponding to the heat of reaction, ΔH, is not constant and increases with increasing partial pressure. Garn [50] deduced Eq. (3.49) for the dependence of ΔH on partial pressure. If the reaction is of the type

$$AB_{(solid\,phase)} \rightarrow A_{(solid\,phase)} + B_{(gas\,phase)} \tag{3.47}$$

then the change in the equilibrium constant with temperature is given by the expression

$$\frac{d \ln K_p}{dT} = \frac{d \ln P_B}{dT} = \frac{\Delta H_0}{RT^2} + \frac{1}{RT^2} \cdot \int_0^T \Delta c_p \, dT \tag{3.48}$$

where ΔH_0 is the enthalpy change of reaction (3.47) at temperature T_0 and P_B is the partial pressure of B.

If the change of Δc_p with temperature is neglected, the expression becomes on integration

$$\ln \frac{P_2}{P_1} = -\frac{\Delta H_0}{R} \left(\frac{1}{T_2} - \frac{1}{T_1} \right) + \frac{\Delta c_p}{R} \ln \frac{T_2}{T_1} \tag{3.49}$$

where P_1, P_2 are the partial pressures of the gas, T_1, T_2 are the temperatures of the peaks, c_p is the specific heat at constant pressure, Δc_p is the difference between the c_p values of the products and the starting compounds, K_p is the thermodynamic equilibrium constant.

On application of this equation to the dehydration of kaolin at heating rates of 6 and 10 °C/min, the conclusion is reached that the partial pressure of the water vapour inside the sample at the peak temperature is approximately four times as great at faster heating rates [3]. The dependence of ΔH on the heating rate is shown graphically in Fig. 3.17. With a slow heating rate, water may be eliminated by diffusion at a rate sufficient to leave a low partial pressure of the water vapour inside the sample.

The sample dehydrates at temperature T_1 and the water liberated absorbs the quantity of heat H_1. Its heat content increases with temperature and at temperature T_2 it has the value H_2. With a higher heating rate, dehydration takes place at a higher temperature T_2 in view of the higher value of the partial pressure. As the product has a higher heat content H_2, the heat of reaction, ΔH_2, is greater than the heat of reaction at the lower temperature.

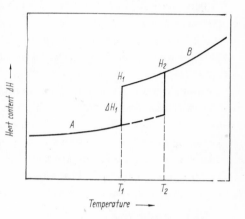

Another factor important in quantitative analysis is the fact that equal amounts of heat produce temperature effects of various magnitudes at various temperatures. In a series of quantitative studies it was demonstrated that when identical amounts of a substance with a known heat of reaction were used, the area

Fig. 3.17 Dependence of heat content on temperature

The part of the curve indicated by A refers to the starting material, that indicated by B refers to the reaction product.
(Reprinted from [47] by permission of the copyright holder)

decreased gradually with increasing temperature. Sabatier [135] demonstrated this fact by comparing standard substances with known heats of reaction which decompose within the temperature interval from 160 to 875 °C (see Table 3.2). This effect was demonstrated clearly by

Table 3.2

Effect of Decomposition Temperature on the Magnitude of the Peak Area at a Known Value of Heat of Reaction (Reprinted from [135] with the permission of the copyright holder)

Substance	Decomposition temperature °C	Heat of reaction cal/g	Peak area, S mV sg^{-1}	Q/S
$CaSO_4 \cdot 2\,H_2O$	160	153	1720	0.089
$ZnCO_3$	425	114	965	0.118
$MgCO_3$	565	320	2160	0.140
$CaCO_3$	875	404	2120	0.190

heating two identical inert samples at a linear rate and inserting into one of them a calibrated heating coil of 5.6 ohm resistance, and introducing an equal amount of heat into the sample at each 100 °C interval (with a heating current of 0.68 A for 3 minutes, this represents 1.100 kcal). The value of the peak area produced in this manner decreased gradually with increasing temperature [42]. For quantitative analysis it follows that calibration of the apparatus (either by standards or calibration heating) should be carried out at a temperature which corresponds to that at which the change under study takes place. Substances used for calibration are discussed in Chapter 4.

Among factors affecting the quantitative evaluation of the curve the frequently employed method of dilution of samples with an inert material should be included. Such dilution can decrease the peak area and it requires the measurements to be carried out at higher sensitivities. This is also true when calibration standards are used. The method of internal standards of known heat capacity is used by Yagfarov [59] for simultaneous determination of heat capacity, thermal conductivity, and heat of reaction.

3.7.2. Methods of quantitative evaluation of curves

It has already been shown that the heat liberated or absorbed during the reaction causes a deviation from the baseline and the peak area is a measure of the heat of reaction. The integral of the effect, i.e. $\int \Delta T \, dt$ within the limits of the beginning of the effect and its return to the baseline expresses quantitatively the magnitude of the heat of reaction and hence also the amount of the active substance. This was explained in Section 3.7. In the literature a number of more or less empirical methods of integration of this area and methods of its graphical solution can be found. The best known are planimetry, cutting and weighing, and counting of squares. It was explained that during the reaction under investigation the fundamental properties of the sample are changed, mainly thermal conductivity, density, and specific heat, (which leads to a change in thermal conductivity and to a change of the deviation of the curve from the baseline). This unfavourable effect may be eliminated by diluting the sample with an inert material, or by separating the sample and the thermocouple, as in the method of Boersma [25] described earlier. The change of the deviation from the baseline causes difficulties in the graphical evaluation of the area, mainly in defining the line limiting the area. The method of evaluation of this line has been developed by a number of workers and is shown in Fig. 3.18. According to Norton [114] the line limiting the area is the straight line drawn between the points of greatest curvature on both sides of the peak (Fig. 3.18a). This method was found suitable for symmetrical peaks only. In the case of an asymmetric peak (Fig. 3.18b) Berkelhamer [16] proposed constructing a line vertical to the tangent at point C, obtaining the triangle BCD as the area required. As under certain circumstances the amount of the reacting material is proportional to the maximum deviation, this deviation is also used for quantitative evaluations [54, 114]. According to Dean [37], in the case of an endothermic effect the cosecant of the angle α formed by the extension of both arms of the peak is proportional to the amount of the active substance (3.18d). According to Mackenzie [99] this method of evaluation is usable only for certain peaks with a constant value of X.Z, or for a symmetrical peak if Y is constant, irrespective of the amount of the active substance. The construction of the reference line is much more difficult in cases where the baseline has been shifted appreciably (3.18e). While in the first case (connecting line AB) no complications from the shift were observed, in the second case (connecting line CDE) Berg [15] recommended constructing the line DF vertical to the baseline (from the point of the peak maximum). The lines CD

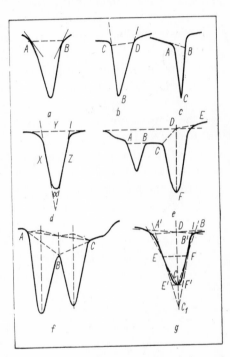

Fig. 3.18 Methods of evaluation of the peak area on a DTA curve

and DE enclose the area to be measured. The problem is even more difficult in a case when two neighbouring effects are superimposed, e.g. in the case of dolomite. Figure 3.18f shows the technique of evaluation according to Berg [15]. In all these cases it is advantageous if the effects of similar shape are compared. For well-shaped endothermal peaks Wittels [163] uses a simple method for the case of a shifted baseline, consisting in an extension of the original baseline (Fig. 3.18c). Krais [72] uses the method of triangle construction for quantitative evaluation of bayerite and böhmite (Fig. 3.18g). The magnitude of the deviation, DC, is determined by drawing a vertical line from the peak intersecting the baseline at C. The area is found by multiplying the height by the width measured at half the height. In cases where the baseline does not make a right angle with the height, it is necessary to construct a straight line perpendicular to the height from their intersection, the so-called baseline AB. This method of evaluation based on the area of the triangle ABC is more accurate than the method of extended sides $(A'B'C')$, especially at low concentrations of active substances. At higher concentrations when an undesirable extension of CC' takes place, the method of two triangles $A'B'C'$ and $E'F'C'$, due to Bárta and co-workers [8] can also be used.

In the evaluation of the temperature effect only that part of the curve which corresponds to the time interval in which the change actually took place is required. The point T_x in Fig. 3.3, corresponding to the end of the effect, is not easily determined. The part of the curve between T_x and T_3 corresponds to temperature compensation which depends on the properties and the construction of the detecting part of the apparatus. Thus only the first part of the curve between T_2 and T_x (Fig. 3.3) is related to the heat of reaction of the change under investigation. This is directly related to the general method of description of the curve and the expression of characteristic points on the temperature axis. In order to define better that part of the curve corresponding to the change occurring, the standardization committee of ICTA recommends defining the curve by three parameters, viz.

(a) the break-point of the peak (indicated usually by T_0), i.e. the point on the curve from which its deviation from the baseline begins;

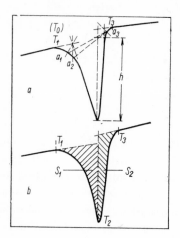

Fig. 3.19 Method of evaluation
of the peak area

(Reprinted from [169] by
permission of the copyright holder)

(b) half-width of the peak, i.e. the distance of the point T_0 from the intersection of the extrapolated baseline with the tangent to the continuation of the curve after point T_0;

(c) temperature of the peak maximum, indicated in this book by T_m [9].

This method of indication will be of importance mainly in the case of recently published results, as it will be impossible to evaluate older papers which do not represent the curve by this method.

According to what has been said concerning the relationship between the area under the curve and the heat of reaction, the area S_1 (see Fig. 3.19) is of the utmost importance. Delimitation of the area by the straight line through points T_1 and T_3 is not correct because then the course of the curve before and after the effect is disregarded [169]. The use of point a_1 (on curve a), which is the transformation point (T_0), was not found suitable. Better results were obtained when a straight line was constructed through points a_2 and a_3, which are the points where the verticals from the points of intersection of the baselines and the continuations of the curve intersect the curve. Almost identical results were obtained when the areas were evaluated according to Fig. 3.19b, where the sum of S_1 and S_2 determines the total area under the curve.

Reproducibility of the curve is a fundamental condition for the measurement of the peak area in quantitative DTA. It depends on the factors which were discussed in detail in Section 3.4. The effect of the composition of the atmosphere in decomposition reactions is a factor of great importance closely related to the rate of heating and the amount of the sample. Equally important factors are the method of preparation of the sample, its packing and shape, which have an influence on the reproducibility of the peak temperature. It is quite clear that for quantitative treatment it is better to use smaller samples with a suitable method of temperature measurement. Large samples, which are often used in DTA, can lead to a number of complications, mainly with respect to their shape; also large temperature gradients are formed, and broad temperature intervals are covered by single peaks. According to Wittels [162, 163], in measuring large heats of reaction, (e.g. $CaCO_3$, 40 kcal/mole) the use of small undiluted samples (3 mg) is suitable if the heating rate is relatively large (30 °C/min). When small thermal changes are measured the use of large samples is also not suitable. The same is true of samples diluted by an inert substance [13, 29].

It may be concluded that DTA curves can be interpreted quantitatively only under carefully chosen experimental conditions guaranteed to give good reproducibility. In decomposition reactions special attention should be devoted to factors which affect the course of the curves, and to the limits of applicability of the method. The basic factors which can change during the experiment and so affect the calibration of the whole system are the sample density ϱ, its thermal conductivity λ, and the coefficient of thermal conductivity of the thermocouple Λ. The rate of heating does not affect the calibration if the thermal conductivity of the sample remains unchanged in the course of the reaction. The greatest errors in quantitative DTA are caused by differences in heat transfer through the sample and the thermocouple. These interfering effects may be reduced to a minimum by standardization of sample pretreatment (reproducible density, volume, and geometrical parameters). The sample is best diluted with an inert material (most commonly alumina), so that the thermal conductivity is determined mainly by the diluent and remains unchanged even when the active substance is sintered or melted. The effect of the thermocouple should be minimized by careful calibration by means of chemical standards with known values of the heats of reaction, and by the use of relatively thin thermocouple wires.

3.8. APPLICATION TO REACTION KINETICS

As in TG, information may be obtained from DTA curves on the mechanism and the kinetics of the process. The general shape of the curve may be used to specify the reaction. In endothermic reactions a pure phase change is characterized by a gradually increasing deviation followed by a sharp peak and a rapid return to the baseline, which may have been shifted. If a diffusion-controlled reaction is occurring, the endothermic effect is usually shallow and the peak is rather rounded. The descending side of an endothermic peak is often described as a straight line. This is hardly ever observed when large samples are used. The shape of the curve does not, of course, depend only on the type of reaction, but may also be affected by experimental factors such as the position of the sample in the block or an isolated crucible, the formation of a temperature gradient, or the change in thermal physical properties of the sample. Kinetic evaluation of DTA curves uses the total area of the curve due to the reaction under investigation, the maximum deviation, representing the greatest reaction rate, and the symmetry. The curve, however, is rarely symmetrical. Merzhanov [103] demonstrated that the curve due to a phase transition is not symmetrical. When different effects are taken into consideration, such a direct symmetry cannot even be expected. The DTA curve is usually characterized by the temperature of the start of the change, the temperature of the point of intersection of the baseline with the extended straight lines of the descending or the ascending sides, the temperature of the peak, the temperature at the maximum rate of reaction, and the temperature at the termination of the reaction. If the temperature is measured in the centre of the sample, then the temperature of the peak corresponds to the termination of the reaction, i.e. to the temperature when the reaction zone reaches the thermocouple junction. In this case the temperature of the peak is the most important temperature point. If the measured temperature is that of the reference sample, the transformation is terminated after the peak, and the important temperature point is that of the beginning of the reaction. The other temperature points are complex functions, as was explained earlier. The area under the curve is not totally related to the course of transformation as the last part is related to the restoration of the temperature equilibrium in the thermocouple system. However, results show that even when the required conditions are imperfectly fulfilled, good agreement is achieved with the results of other methods of evaluation of kinetic data.

The methods of investigating reaction kinetics by DTA were reviewed by Šatava [151], Wendlandt [160, 161], and recently by Šesták and Berggren [169].

The first use of DTA in the study of reaction kinetics was made by Murray and White [105]. They studied the relationship between the rate of decomposition of kaolin and temperature (rate of heating, 10 °C/min); this required a knowledge of the weight loss as a function of time at constant temperature. Thus, they obtained a trace similar to a DTA curve and showed that at the peak temperature decomposition was only 74%. The curves, calculated from the rate constants obtained thermogravimetrically, are similar to the DTA curves. They show the effect of the heating rate on the temperature of the peak. The authors found that dehydrations follow first-order kinetics,

$$\frac{dx}{dt} = Z \cdot e^{-E/RT} \cdot (100 - x) \tag{3.50}$$

where x is the percentage decomposed in time t. They determined the peak temperature, which they took as the temperature at which the reaction rate was a maximum. They assumed a homogeneous distribution of temperature which, in fact, is not the case. The curves calculated from the rate constants obtained by isothermal weight change determination display a certain agreement with the DTA curves. For example, an identical shift of the peak temperature to higher values with increasing heating rate (5, 10 and 20 °C/min) can be observed, but the magnitude of the deviation increases with decreasing heating rate.

From the relationships deduced, Murray and White calculated the temperature of the peak on the DTA curve corresponding to dehydration, for the usual rates of heating, i.e. 10 and 20 /C°min. They obtained the condition for a maximum by differentiating Eq. (3.50), and setting equal to zero, viz.

$$Ze^{-E/RT_m}\left(-\frac{dx}{dt}\right) + (100 - x)\,Ze^{-E/RT_m} \cdot \frac{E}{RT_m^2} = 0 \tag{3.51}$$

into which they substituted the expression for dx/dt from relationship (3.52) for the heating rate of 10 °C/min:

$$\frac{dx}{dt} = \frac{dx}{dT} \cdot \frac{dT}{dt} = \frac{dx}{dT} \cdot \frac{10}{60} \tag{3.52}$$

Substituting in (3.51) gives

$$-6Ze^{-E/RT_m} \cdot \frac{dx}{dt} + (100 - x)Ze^{-E/RT_m} \cdot \frac{E}{RT^2} = 0 \tag{3.53}$$

the solution of which gives the following expression for the temperature of the peak:

$$\frac{E}{RT_m^2} - 6Ze^{-E/RT_m} = 0 \tag{3.54}$$

or, if the heating rate was 20 °C/min:

$$\frac{E}{RT_m^2} - 3Ze^{-E/RT_m} = 0 \tag{3.55}$$

Allison [2] also studied the kinetics of the dehydration of kaolinite and used the method of Vold [154]. When determining the reaction order he used the relationship

$$\frac{dx}{dt} = k(1 - x)^n \tag{3.56}$$

where dx/dt is the rate of change, x is the amount of substance transformed in time t, k is the rate constant, and n is the reaction order. From the relationship between $\ln k$ and $1/T$, which is linear when $n = 1$, he determined the reaction order.

Kissinger [68, 69] developed a method of measuring the activation energy E and the frequency factor Z from the DTA curves for unimolecular reactions and he found that the temperature and shape of the peak are dependent on the heating rate and may be used for the determination of the kinetics of the reaction. He started from the fundamental kinetic equation:

$$-\frac{1}{V} \cdot \frac{dx}{dt} = Ze^{-E/RT} \cdot \left(\frac{x}{V}\right)^n \tag{3.57}$$

where E is the activation energy, x is the number of particles of the starting substance, n is the reaction order, Z is the frequency factor, and V is the volume.

This equation applies for any temperature T, variable or constant, if T and x are measured at the same moment. If the temperature increases at constant rate dT/dt the reaction rate increases to a maximum when $d/dt \cdot (dx/dt) = 0$. Hence, from Eq. (3.57) it follows that

$$\frac{d}{dt} \cdot \left(\frac{dx}{dt}\right) = \frac{dx}{dt}\left[\frac{E \cdot \dfrac{dT}{dt}}{RT^2} - Zn\left(\frac{x}{V}\right)^{n-1} \cdot e^{-E/RT}\right] = 0 \tag{3.58}$$

and further

$$\frac{E \cdot \dfrac{\mathrm{d}T}{\mathrm{d}t}}{R \cdot T_m^2} = Zn \cdot \left(\frac{x}{V}\right)_m^{n-1} \cdot e^{-E/RT_m} \tag{3.59}$$

T_m is the temperature at which the reaction rate is greatest. The expression $(x/V)_m$ is the amount of substance which remains undecomposed at the moment of maximum reaction rate. By integrating Eq. (3.57) and solving Eq. (3.59) for $n = 0$ and $n = 1$ Kissinger obtained

$$n\left(\frac{x}{V}\right)_m^{n-1} = 1 + (n-1)\frac{2RT_m}{E} \tag{3.60}$$

in which the expression $2RT_m/E$ has a value of less than unity; if the dependence of T_m on the heating rate is neglected, then $n(x/V)_m^{n-1}$ is also independent of heating rate and is approximately equal to 1. After substitution into Eq. (3.59), taking logarithms and differentiating, the following equation is obtained:

$$\mathrm{d}\ln\frac{\dfrac{\mathrm{d}T}{\mathrm{d}t}}{T_m^2} = -\frac{E}{R} \cdot \mathrm{d}\frac{1}{T_m} \tag{3.61}$$

i.e.

$$-\frac{E}{R} = \frac{\mathrm{d}\ln\left(\dfrac{\mathrm{d}T}{\mathrm{d}t}\Big/ T_m^2\right)}{\mathrm{d}\left(\dfrac{1}{T}\right)} \tag{3.62}$$

In practice, several DTA curves are recorded at the same sensitivity with the same weight of sample, but at different heating rates, $\mathrm{d}T/\mathrm{d}t$. T_m values are then determined and a graph of $\ln\left(\dfrac{\mathrm{d}T}{\mathrm{d}t}\Big/ T_m^2\right)$ vs. $1/T_m$ is plotted. A straight line is thus obtained, the slope of which is $-E/R$ and hence the activation energy, E, may be calculated. The value of the frequency factor can be calculated from Eq. (3.59) after substituting the calculated value of the activation energy, assuming that $n(x/V)_m^{n-1} \to 1$.

For the deduction of the reaction order Kissinger made use of the formal kinetic equation which is valid only in cases when $n = 0$, 1, 1/2, and 2/3 (i.e. unidirectional diffusion, advanced state of the growth of nuclei, movement of the reaction boundary from the surface to the interior of the particles). The relationship between the shape of the DTA curve and the reaction order is shown in Fig. 3.20.

As the value of the reaction order decreases, the curves become less symmetrical (at constant heating rate, E, and Z). The change of the reaction order does not affect the temperature of the peak, T_m, but only the shape of the curve. The values of the activation energy and the frequency factor

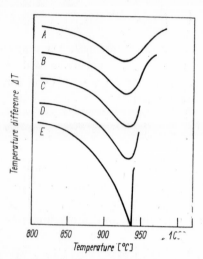

Fig. 3.20 DTA curves of dehydration of kaolinite showing the effect of the reaction order on the shape. The curves were calculated on the assumption that the heating rate was constant and that the frequency factor and the activation energy were unchanged.

$A - n = 1$; $B - n = 3/4$; $C - n = 1/2$;
$D - n = 1/3$; $E - n = 0$. (Reprinted from [69] by permission of the copyright holder)

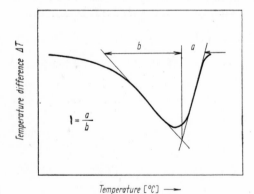

Fig. 3.21 Method of determination of the symmetry of the temperature effect according to Kissinger. (Reprinted from [69] by permission of the copyright holder)

also have a relatively weak influence on the shape of the curve. To express the asymmetric form of the curve quantitatively Kissinger introduced the "shape index of the DTA curve", denoted by I, by which he expresses the shape of the curve and its dependence on the value of the reaction order. The evaluation of this index is shown in Fig. 3.21.

The curve is divided into two triangles by means of tangents to both sides. The value I is given by the expression

$$I = \frac{a}{b} = \frac{(d^2x/dt^2)_1}{(d^2x/dt^2)_2} \quad (3.63)$$

where $(d^2x/dt^2)_{1,2}$ holds for values of x and t for which $d^3x/dt^3 = 0$. Kissinger deduced a relationship between the index I and the reaction order:

$$I = 0.63 . n^2 \quad (3.64)$$

i.e.

$$n = 1.26 . I^{1/2} \quad (3.65)$$

Thus, after computation of I from the DTA curve, the reaction order may be calculated. Kissinger checked his method by a study of the decomposition of kaolinite and found that his values of E and Z were in good agreement with results obtained by Murray an d White by the isotherma method. Kissinger's method does not take into account any geometrical factors in the heating of the sample and, therefore, is an approximate method, primarily for first-order reactions. It should also be kept in mind that physical properties of the sample may possibly shift the peak temperature, leading to erroneous results for the values of the activation energies measured in this manner.

Kissinger's method served as the basis used by Piloyan and his co-workers [122, 123]. However, whereas Kissinger always obtained several DTA curves at various heating rates, Piloyan proposed determining the activation energy E from a single curve recorded at an arbitrary heating rate, when the decomposition of solid substances was followed. The substance under investigation decomposes within a given temperature range and produces a peak in which the initial deviation from the baseline is given by the relationship:

$$\Delta T = B \frac{d\alpha}{dt} \tag{3.66}$$

where ΔT is the deviation of the point of maximum curvature from the baseline (°C), B the area under the curve (°C . sec), α the degree of transformation and $d\alpha/dt$ the reaction rate. Piloyan did not determine the reaction order and concentrated on the determination of activation energy. As the basis for this he used Eq. (3.67):

$$\frac{d\alpha}{dt} = Z \cdot f(\alpha) e^{-E/RT} \tag{3.67}$$

where Z is the constant from the Arrhenius equation $(k = Ze^{-E/RT})$, $f(\alpha)$ is a function of the degree of transformation [it is often assumed that $f(\alpha) = (1 - \alpha)^n$ where n is the reaction order; generally, however, $f(\alpha) = \alpha^m \cdot (1 - \alpha)^n$, where m and n are constants]. Combining (3.66) and (3.67) and taking logarithms:

$$\ln \Delta T = \ln \left[Zf(\alpha)/B \right] - E/RT = C + \ln f(\alpha) - E/RT \tag{3.68}$$

where C is a constant and T is the temperature (expressed in K) of the point of maximum curvature.

Under normal conditions the only variable determining the reaction rate is α. However, in DTA the temperature also changes, and at common heating rates (10−45 °C/min) the value of α is within the limits 0.05−0.08. When solving Eq. (3.68) the expression $\ln f(\alpha)$ is neglected and E determined directly from the relationship

$$\ln \Delta T = C - E/RT \tag{3.69}$$

where the ΔT value is determined directly from the DTA curve. The maximum rate of transformation can be determined from

$$\left(\frac{d\alpha}{dt}\right)_m = \frac{\Delta T_m}{\Delta B} \tag{3.70}$$

When E and Z are known the rate constant may be calculated. The method was applied to the dehydration of $CuSO_4 \cdot 5 H_2O$ and $CaC_2O_4 \cdot H_2O$, and to the decomposition of $MgCO_3$ with good results.

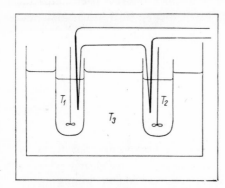

Fig. 3.22 DTA apparatus for the evaluation of kinetic data with a homogeneous medium (solution) according to Borchardt and Daniels. (Reprinted from [27] by permission of the copyright holder) T_1, T_2, T_3; Temperature of homogeneous media.

A study of reaction kinetics by means of DTA was carried out by Borchardt and Daniels [27] with homogeneous reaction mixtures and solutions. They worked with solutions and the construction of their apparatus is shown in Fig. 3.22. Stirrers are used to prevent the formation of temperature gradients and the temperature difference is measured by a differential thermocouple. The authors deduced an equation for the calculation of the rate constant, based on the following assumptions: (a) the temperature in both vessels is the same, (b) heat is transferred only by conduction (the heat transfer by the thermocouple is disregarded), (c) the coefficient of thermal conductivity and heat capacity of the sample and the standard are identical, (d) the reaction rate is negligible at the starting temperature of the measurement, (e) the reaction enthalpy is sufficiently large, (f) the liberated heat is proportional to the number of moles reacted, (g) activation energy, reaction enthalpy, heat capacity of the solution, and the coefficient of thermal conductivity are temperature-independent in the given temperature range. Some of these conditions can only be fulfilled by using solutions. The following relationships were deduced for the rate constant:

$$k = \left(\frac{KSV}{q_0}\right)^{m+e-1} \cdot \frac{\left[\dfrac{c\,d\Delta T}{dt} + K\Delta T\right]}{[K(B-a)-c\Delta T]^e \left[K\left(\dfrac{M_0}{L_0}\varrho - \dfrac{m}{l}\cdot a\right) - \dfrac{mc\Delta T}{l}\right]^n} \tag{3.71}$$

$$k = Ze^{-E/RT} = \left(\frac{KSV}{u_0}\right)^{n-1} \cdot \frac{\left[c\dfrac{\mathrm{d}\Delta T}{\mathrm{d}t} + K\Delta T\right]}{[K(B-a) - c\Delta T]^n} \qquad (3.72)$$

where L_0 and M_0 are the starting numbers of moles of substances L and M, m and l are the numbers of moles of substances L and M, u_0 is the number of moles of the starting substance, K is a constant characteristic of the given apparatus, which is determined from $\Delta H = KB$, B is the area under the curve (min. °C), a the area under the curve up to moment t, c the heat capacity of the solution, d $\Delta T/\mathrm{d}t$ the gradient of the tangent at point t (°C/min), and n is the reaction order.

If the conditions mentioned above are fulfilled, it is possible to estimate graphically from the DTA curve all the values necessary for substitution into the equations. Only the reaction order remains unknown and the authors determined it by trial and error. If the reaction order is chosen correctly the plot of ln k versus $1/T$ is linear. The gradient of this straight line represents the value of E/R. This method was checked in studies of the decomposition of benzenediazonium chloride and of the reaction of ethyl iodide with N, N-dimethylaniline in aqueous solutions. This method was not found suitable for the study of solid phase reactions [1] and would require further investigation. For the investigation of heterogeneous reactions Borchardt's method was used by Blumberg [23] who studied the reaction between SiO_2 and hydrogen fluoride.

The disadvantage of Borchardt's method is the tedious process for finding the reaction order. Freeman and Carroll [46] showed that this procedure may be simplified by using the following relationship:

$$\frac{\Delta \ln\left[c\dfrac{\mathrm{d}\Delta T}{\mathrm{d}t} + K\Delta T\right]}{\Delta \ln[K(B-a) - c\Delta T]} = -\frac{E}{R}\frac{\Delta 1/T}{\Delta \ln[K(B-a) - c\Delta T]} + n \qquad (3.73)$$

If the left-hand side of the expression is plotted against $(\Delta 1/T)/\Delta \ln[K(B-a) - c\Delta T]$ a straight line is obtained, the slope of which gives the value E/R ($E/2.303R$ if Briggs logarithms are used); the order of reaction is given by the intercept on the ordinate. However, application of this method for the determination of the reaction order did not produce reproducible results, and therefore some authors discourage its use.

Reich [127] applied Borchardt's deductions to solid substances for the determination of kinetic constants from DTA curves.

Methods of using DTA or differential calorimetry (see Sections 3.9.5 and 3.9.6) to study reaction kinetics are constantly being developed. Recently Taylor and Watson [175] used differential scanning calorimetry for the determination of relative reaction rates, by comparison of two analogous curves. Reed, Weber and Gottfried [176] studied the methods published and concluded that only the Borchardt and Daniels method expresses the kinetics of the reaction with sufficient accuracy. Kissinger's assumption that maximum reaction rate corresponds to the point T_m is not invariably true, and the good results often obtained by Kissinger's method are due to the cancelling of errors as a result of temperature differences inside the sample.

In conclusion it should be said that the preceding section does not deal with all possible methods or procedures for the calculation of kinetic parameters. Only some of the most interesting methods of solution were discussed to show the possibility of applications of this method in this field. In applying the method, however, great attention should be devoted to all parameters affecting the curve, primarily the effects of the composition of the atmosphere used. Reversible reactions are especially sensitive to the gaseous products liberated, and irreversible ones to access of an active gaseous medium. The determination of kinetic data from DTA curves is not yet widespread, and the author considers that investigation of reaction mechanism and a correct interpretation of reaction kinetics from DTA results is very difficult. This is because the method itself creates a temperature difference between the sample and the standard, making the objective measurement of temperature and the determination of the degree of the change impossible. Compared with thermogravimetry the DTA method is less important for studying reaction kinetics. However, these limitations do not apply in its quantitative application, and in this field the author considers that it has great prospects.

3.9. SPECIAL PROCEDURES AND RELATED METHODS

Differential thermal analysis is used and applied in a number of fields for the solution of various problems in thermal behaviour and the analysis of inorganic and organic substances, both in the solid and liquid state. This has contributed to the wide use of the method, very often in special cases where other analytical methods cannot be used. For such methods new procedures have been developed, and will now be discussed.

3.9.1. Derivative DTA

DTA curves, especially for decomposition reactions, do not deviate from the baseline at a sharply defined temperature, but are characterized by a gradual deviation. In quantitative work the area between the curve and the baseline must be determined accurately, which is difficult in such cases. The derivative DTA curve is of value in this case, consisting of a plot of $d(\Delta T)/dt$ versus temperature or time (where ΔT is the temperature difference). The curve obtained in this manner shows small changes in the normal DTA curve better; further, it makes some effects clearer, and permits a more accurate determination of the baseline, which is very important in quantitative work and in investigations of reactions kinetics. When a normal DTA curve is compared with a derivative DTA curve it can be seen that on the latter a positive and negative double peak corresponds to a simple DTA peak. At T_m the derived curve intersects the zero line and its maxima or minima correspond to the inflexion points on the normal DTA curve. This method is useful in that it resolves small, partly superimposed effects, which are much clearer on the derivative DTA curve. This method was first used by Burgess [30]. Freeman and Edelman [45] used a physical method to obtain the derivative DTA curve, employing two identical reference materials with a constant temperature difference maintained between them. This

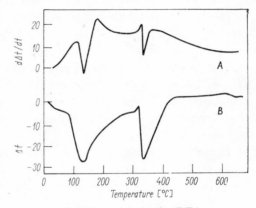

Fig. 3.23 DTA and derivative DTA curves for a sample of KNO_3.

A — DTA curve. B — Derivative DTA curve. (Reprinted from [45] by permission of the copyright holder)

method is an analogy of the technique of differential thermogravimetry according to de Keyser [65]. The thermocouple in one sample was situated approximately 3 mm higher than that in the other, and the samples were arranged asymmetrically with respect to the heating coil. The constant temperature difference between the two samples was 3.3 °C over the whole temperature interval of 100 − 700 °C. A typical derivative DTA curve is shown in Fig. 3.23. The minima on this curve correspond to the maximum change in rate of the temperature difference. The derivative curve may be obtained not only with an asymmetrical position of the samples and the thermocouples in the furnace (in order to achieve a constant temperature difference, as mentioned), but also by Keyser's method, i.e. by using two identical furnaces and thermocouple systems, and keeping the rate of heating constant with the temperature in one furnace higher by a constant amount. This temperature difference should be suitably chosen. If it is too large, two DTA curves will be obtained, and if the difference is too small, a derivative DTA curve will be obtained but with two shallow peaks. Currently, the derivative DTA curve is usually obtained by an electrical method. Campbell and co-workers [32] used the electric method for the derivation of the curve and deduced from their measurements that the height of the corresponding peak on the derivative DTA curve is, in the case of a phase change, proportional to the concentration of the transformed component. Frederickson [44] carried out a manual differentiation of the DTA curves of kaolinite with the aim of a better interpretation of the results obtained. His procedure was to evaluate temperature differences, ΔT, for successive identical temperature intervals $\Delta T' = 10\ °C$. By plotting the expression $\Delta T/\Delta T'$ versus T' he obtained the derivative curve.

A certain modification of this method is called double differential thermal analysis in which the reactive sample is compared with a reactive reference sample. The aim of this comparison is to compensate identical reactions at identical temperatures, thus obtaining a smooth line, or, by suitably choosing the active reference substance, to eliminate only some effects and increase the resolving power of others. This method, developed by McLaughlin [77], presents many difficulties, mainly in the requirement of identical heating of both samples, and is important only in special cases, for example in investigation of small changes of chemical composition of two otherwise identical substances.

3.9.2. Boersma's method of synthesis of DTA curves

As was shown in Section 3.3, Boersma [25] discusses certain disadvantages of the current DTA method and proposes a new measuring technique,

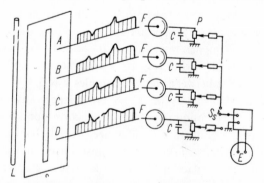

Fig. 3.24 Synthesis of DTA curves according to Boersma.

A, B, C, D — moving templates; E — oscillograph; L — lamps; S — slit; F — photocells; C — integrating condensers; P — calibrated potentiometers; S_S — selector switch.
(Reprinted from [25] by permission of the copyright holder)

consisting in the use of a new sample holder (see Chapter 5, Fig. 5.1) which he uses in differential calorimetry. He also proposes a new method of analysing the curves, the so-called "synthesis of the curve". This technique consists in adding together the curves of the pure minerals present in the given substance and comparing the result with the curve obtained for the mineral substance itself. The curve is synthesized by using photocell equipment and is transferred to an oscilloscope, where it is compared with the curve of the analysed substance. The principle of this method is shown in Fig. 3.24. The method is original but in practice it is limited because it does not take into account factors affecting characteristic temperatures, such as mutual effects of the mixture components, or the influence of atmosphere, etc. The method is rather complex experimentally and it appears to be not very suitable for quantitative analysis.

3.9.3. DTA as a microanalytical method

In current DTA the sample weight ranges from ten to several hundred milligrams. This is called the macro-method. In contrast to this the micro-method requires only several milligrams. The microanalytical method in DTA has been used only over the past ten years. In the current macro-analytical method a number of factors affecting the determination should be taken into consideration. These have been enumerated and discussed in the preceding sections. It is due to these factors that high accuracy and reproducibility of the results cannot be expected in this method, and it follows that the applications are often limited, especially in quantitative work. These problems were discussed in 1967 at the first International Conference on Thermal Analysis and a standard method of publication of the results of thermal analysis was recommended [9, 101]. The main reason for the non-reproducibility in current macro-methods is the use of an appreciable amount of the sample, causing a non-uniform distribution of temperature in the sample, non-uniform liberation of gases, and a number of other effects connected with thermal conductivity, porosity, degree of packing, etc. The majority of these problems connected with the macro-method can be eliminated by decreasing the sample size. Since 1961 the number of attempts at miniaturization has steadily inceased. This miniaturization is not an end in itself, but ensures better control of experimental conditions. The decrease in the sample weight to the microscale (tenths to tens of milligrams) ensures a better homogeneity of the temperature distribution within the sample, a greater accuracy in temperature measurement, and an increased resolving power and reproducibility of the transformations investigated. The quantitativeness of the method is improved to a degree which is only achieved with difficulty by the macro-method. The micro-method is better suited for quantitative purposes, as quantitative evaluation of the DTA curves requires mainly that the area under the curve is independent of the thermal conductivity of the sample, i.e. that the same heats of reaction produce identical areas, indenpendent of the state of the sample. The decrease of temperature gradients and the uniform distribution of the temperature in the sample lead to the

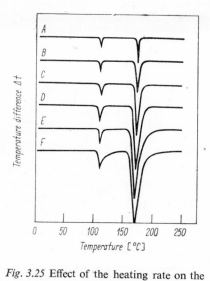

Fig. 3.25 Effect of the heating rate on the melting point of hexamethylbenzene in micro DTA. Sample weight 3.4 mg; recorded on a Shimadzu DTA—20 B apparatus.

Rates of heating in °C/min: $A = 2$, $B = 5$, $C = 10$, $D = 15$, $E = 20$, $F = 30$.

(Reprinted from [150] by permission of the copyright holder)

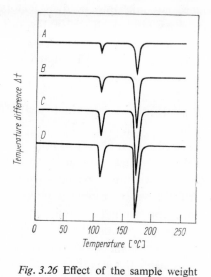

Fig. 3.26 Effect of the sample weight on the melting point of hexamethylbenzene at 10 °C/min heating rate; recorded on Shimadzu DTA—20 B apparatus.

A — 0.7 mg; B — 1.4 mg; C — 2.1 mg; D — 3.4 mg.

(Reprinted from [150] by permission of the copyright holder)

temperatures of the beginning and the maximum of the effect peak being independent of the heating rate, as well as of the sample weight (within the range given). This is demonstrated by Figs. 3.25 and 3.26 which were obtained with a sample weighing less than 3.7 mg. Miniaturization permits a substantial decrease in the thermal capacity of the furnace and an improved accuracy in the control of the temperature; this increases the sensitivity and the resolving power of the method to such a degree that the thermal history of the sample is detectable, e.g. in glass and polymers. It also enables DTA to be used in the study of phase diagrams.

Micro DTA also has certain attendant problems, mainly with respect to the apparatus, such as the requirement of thermal homogeneity of the reference and the analysed samples, low heat capacity and conductivity of the sample holders, thermal insulation of the sample, and an accurate determination of the position of both thermocouples to give equivalence under identical conditions of heating or cooling. The technique using milligram amounts of the sample requires, naturally, a higher amplification of the differential signal from the thermocouples, which necessitates greater care in the manufacture of transducers. The increase in sensitivity may be achieved to a certain extent by adapting the method to the technique of differential calorimetry, i.e. by insulation of the sample holders; however, this should be done only to an extent which would not impair the resolving power of the method. In this respect the construction of the holders is already satisfactory.

In principle the problem of the micro-method may be dealt with in two ways. First by layering the active and the inert material, and placing the thermocouple junction in the layer of the active substance (see Fig. 3.16) or mixing a small amount of the active substance under consideration with the inert material and employing the usual technique with the homogeneous mixture obtained. However, this procedure requires a very sensitive apparatus and does not give good results and it is more convenient if the active substance is concentrated between two layers of inert material. According to Barshad and other authors the magnitude and the clarity of the peak depend on the concentration of the investigated substance at the thermocouple junction. A second possible method is that used by Mazières [100], shown in Fig. 3.27. Here the thermocouple junction serves as the sample holder, and its shape, as well as the size of the cavity, may be modified in various ways. This method is ideal from the point of view of heat flow, the exchange of heat not being complicated by transitions through various boundaries between the block and the crucible, but the removal of heat by the thermocouple wires should be taken into account. The difficulties will be predominantly technical ones, such as working with microquanties of sample, and cleaning of the hollows

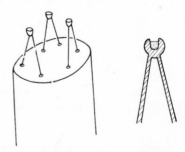

Fig. 3.27 Special crucibles for micro DTA. (Reprinted from [100] by permission of the copyright holder)

in the thermocouple junctions. It has been possible to analyse samples weighing only some tens of micrograms by this method.

It should be mentioned that miniaturization of DTA is not always due to the advantages mentioned, but is sometimes required by the problem under investigation, e.g. Rogers [130] used this method for the investigation of explosive substances, which could not be analysed by a normal technique. At the moment the number of papers dealing with the development and applications of micro DTA is small. All relevant papers are quoted here as the author believes that because of the advantages mentioned the method will play an ever more important role in the future [12, 24, 36, 39, 51, 58, 66, 71, 100, 153, 163, 164, 165, 172]. Miniaturization is not limited to DTA but is also used in thermogravimetry. Although here methods using a microbalance have a much longer tradition, it is only recently that thermobalances of this type have become commercially available (see Chapter 5).

In view of the fact that the development of micro DTA represents one of the major contributions in the field of thermal analysis and that it will assume an important practical role in the future, a detailed list of references in this field is attached to this book as Appendix No. III.

3.9.4. Dynamic differential calorimetry*

The effect of the geometry and the material of the block on the results was discussed in Sections 3.4.2 and 3.7, and it was stressed that in classical DTA, the use of a block allows heat transfer between samples, and thus a rapid equalization of temperature in a relatively short time, making the differentiation of consecutive reactions possible. However, if both samples are thermally insulated from each other, conditions are now similar to those of calorimetry, under which discrimination of neighbouring reactions is impaired. The method, called "dynamic differential calorimetry" (DDC), was developed by Schwiete and Ziegler [139], and gives easy quantitative evaluation of the results. The method is also based on Le Chatelier's differential method. The distribution of temperature inside the sample is not important here, because the measurement is carried out over the whole range of the temperature effect, thus ensuring good reproducibility. The basic arrangement for DDC is shown in Fig. 3.28.

---

* This technique is merely a specialized application of DTA, and as such the term dynamic differential calorimetry was rejected by the ICTA nomenclature committee (see *Talanta*, **16**, 1227 1969). However, for completeness the limited amount of work on this technique that has been published is dealt with here.

Heat changes in the sample cause temperature changes in the crucibles. As can be seen from Fig. 3.28, a certain similarity exists between the DDC arrangement and Boersma's arrangement [25] shown in Fig. 5.1 in Chapter 5, the main difference being that, in Boersma's arrangement, the temperature difference is measured in the region of maximum heat transfer. On the basis of thermodynamic considerations Schwiete deduced relationships for the quantitative evaluation of the temperature effects obtained in this manner, and for the calibration of the equipment. Every change accompanied by a change in heat content is evident as a temperature difference between the calorimeters. This difference is measured by a heat flow ΔI from a heat bath of relatively large capacity which compensates the heat loss created. At a given temperature the temperature difference is determined by the heat content of this body. The integral of the heat flow within the limits t_i and t_e is proportional to the integral of the temperature difference ΔT between these points. When the factor of heat transfer, for the given range of the reaction, is introduced, then equation (3.74) is obtained:

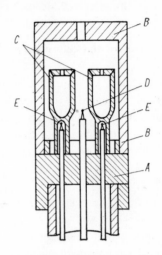

Fig. 3.28 Block diagram of a dynamic differential calorimeter
A — block, B — lid, C — crucibles with lids, D — thermocouple for temperature measurement, E — differential thermocouple.
(Reprinted from [139] by permission of the copyright holder)

$$K\Phi(T) \int_{t_i}^{t_e} \Delta T(t)\,dt = \Delta q = \int_{t_i}^{t_e} \Delta I(t)\,dt \qquad (3.74)$$

For a given geometry the values for $K\Phi(T)$ may be determined from reactions with known heats of reaction. Within a certain range of sample weights, which may be determined from the changes of the total enthalpy of the system as a function of sample weight, the integral on the left-hand side of the equation, i.e. the magnitude of the effect, is directly proportional to the sample weight. The method was checked in the determination of the heats of transformation and fusion of Li_2SO_4, and a number of other reactions were discussed. The advantages of this method compared with DTA are summarized below.

(a) Possibility of quantitative evaluation of the heats of reaction.

(b) The thermocouples are not in direct contact with the sample.

(c) During the subsequent reactions the experiment may be interrupted and the intermediate products identified.

3.9.5. Differential calorimetry

In Section 3.7 it was shown that in quantitative DTA, enthalpy changes may be evaluated when the temperature effects of changes with known values of the heat of reaction or of the heat of transition are used. The results are not measured directly in calories, and the DTA curves require calibration and the areas are evaluated on the basis of calibration values. As these methods are elaborate and inaccurate, modified methods have been developed which give reaction enthalpies and specific heats directly in calories. The basic procedure in these methods remains the same as in classical DTA, but either one or both samples, or the block, contains auxiliary heating elements. These compensate, by means of an accurately measured quantity of electricity, the temperature differences between the samples, due to the transformations occurring. The electrical energy introduced in this way corresponds to the heat of reaction or transformation and is measured directly in calories. Apparatus of this nature, known as a differential calorimeter, is produced commercially. The heating element used is usually made of platinum or a similar metal, and can be placed directly in the analysed sample. Procedures are known where the sample and the inert material are mixed with pure graphite which is heated by passing an electric current. Although perfect contact is achieved between the heating element and the substance, in many cases this method cannot be used because of the special experimental conditions required. The location of the heating element in the sample or reference, or in both, depends on whether the aim is to compensate the exo- or the endothermic effects.

A modification of this method is that of differential scanning calorimetry (DSC) described in the following section. In this method only the sample holder, on which thin-walled metallic pans containing the samples are placed, is subjected to auxiliary heating.

3.9.6. Differential scanning calorimetry (DSC)

Another method based on the original principle of DTA and giving quantitative calorimetric results is differential scanning calorimetry

(DSC) which might also be called inverse DTA. In this method the sample is submitted to linear heating, and the rate of heat flow in the sample, proportional to the instantaneous specific heat, is measured continuously [115, 116, 156]. Inside the measuring cell, which is normally kept at room temperature, two symmetrical pans are fixed. The platinum resistance thermometer and the heating unit built into the sample holders serve as a primary temperature control of the system. The secondary temperature control system maintains the temperature difference between the two holders at zero by means of a heating current which is measured. In other words the temperature of the sample is kept the same as that of the reference sample (or the block) by introducing heat into the reference sample. This amount of heat is recorded as a function of time or temperature. Hence, what is measured is not the temperature difference as in classical DTA, but the electrical input necessary for the maintenance of isothermal conditions. The use of small samples (milligrams) placed on metallic foils reduces the temperature drop to a minimum. A low heat capacity of the whole system permits high heating rates (up to 80 °C/min)

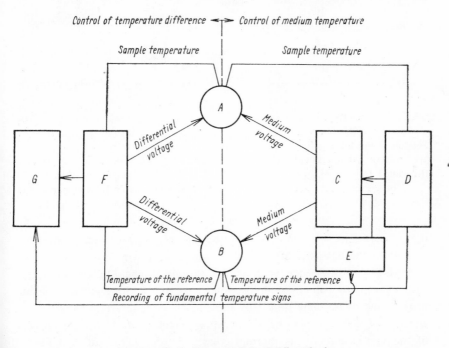

Fig. 3.29 Block diagram of the DSC method

A — sample holder, B — reference holder (empty), C — amplifier of the average temperature, D — average temperature computer, E — temperature programmer, F — amplifier of the temperature difference, G — recorder.

and gives a high resolution. The basic scheme of DSC is shown in Fig. 3.29. The apparatus itself, which at present is produced by two companies (Perkin-Elmer and DuPont) is described in detail in Chapter 5. DSC may be used for direct determination of specific heats, as was shown by O'Neill [116] with samples of graphite, diamond, gold, silver nitrate, etc. The temperature of the sample carrier is programmed as in Fig. 3.30 during measurement. The baseline is obtained from an experiment with no sample. The small deviation from the baseline at the start of the temperature programme expresses the inequality of the heat capacities of

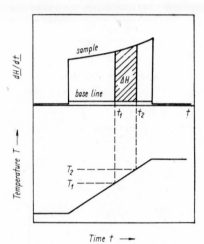

Fig. 3.30 Plot of dH/dt vs. t and T
vs. t for DSC. The base line is
obtained from an experiment carried
out without sample.
(Reprinted from [116] by permission
of the copyright holder)

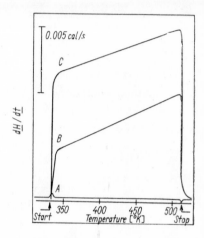

Fig. 3.31 A curve obtained by the DSC
technique (for the determination of
the specific heat of diamond).
A — baseline, B — calibration,
C — sample analysis.
(Reprinted from [116] by permission
of the copyright holder)

the carriers and their contents. At the end of the programme the curve returns to the baseline. Measurement with a known amount of sample is carried out in the same way. If the sample is enclosed in the carrier each element of volume of the sample must follow the given temperature programme dT/dt, determined by the carrier, after the termination of the effect. The duration of the temperature effect depends on the amount of the sample and on its thermal conductivity. The rate of heat flow in the sample is expressed by the equation

$$\frac{\mathrm{d}H}{\mathrm{d}t} = M \cdot c_\mathrm{p} \frac{\mathrm{d}T}{\mathrm{d}t} \tag{3.75}$$

where dH/dt is the rate of heat flow (cal/sec), c_p the specific heat (cal . g^{-1} . K^{-1}), dT/dt is the rate of change of temperature (K/sec), and M is the mass of the sample (g). The area under the dH/dt curve between two temperatures corresponds to the change in the heat content, i.e.

$$H(T_2) - H(T_1) = \int\limits_{t_1}^{t_2} \frac{dH}{dt}\, dt \qquad (3.76)$$

The apparatus is calibrated by using a material of known specific heat, usually α-Al$_2$O$_3$ or synthetic sapphire for which the specific heat has been determined with high accuracy in the temperature range $0-1200$ K. A typical DSC curve for the determination of specific heat of diamond by this method is shown in Fig. 3.31.

The DSC method is being used more often mainly because of the experimental simplicity and availability of the apparatus. For example, Plato and Glasgow [124] used this method to determine heats of fusion (ΔH_f) and changes in them caused by the presence of impurities. They deduced a simple expression for the purity:

$$\text{Mole \% of impurities} = 100 . (\Delta H_f/RT_0^2) . \Delta T \qquad (3.77)$$

where ΔH_f is the heat of fusion, ΔT is the shift of the melting point due to impurities, T_0 is the melting point of pure sample. As a standard they used very pure indium (m.p. 429.7 K, $\Delta H_f = 780$ cal/mole). Using this method they investigated 95 organic compounds.

Foltz and McKinney [43] modified the DSC method by using as reference sample the same compound as the analysed sample, the only difference being

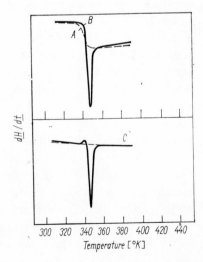

Fig. 3.32 A curve obtained by the modified DSC method.

A — normal curve of an amorphous polymer heated to glass-state temperature; B — curve corresponding to a sample solidified by freezing; the superimposition of both curves expresses the change in c_p for the given transition; C — curve obtained by the modified method, using a simultaneous heating of both preceding samples; the complication connected with the increase in c_p is eliminated. (Reprinted from [43] by permission of the copyright holder) that the reference had been treated differently, giving it special properties. Thus, small energy differences were measured which in ordinary DSC appear as small inflexion points on the curve. They applied the modified method to quantitative observation of the hardening process in polyvinyl chloride in the neighbourhood of the temperature of glass formation (see Fig. 3.32). In this figure an example of an ordinary DSC curve is shown, obtained by using the classical reference material, and a curve obtained by the modified method.

The only disadvantage of present DSC (as met in commercial apparatus) is that it can be used only up to 600 °C. The use of the method at higher temperatures would be advantageous as the determination of specific heats and heats of reaction is relatively simple and direct.

3.9.7. Calvet's microcalorimetric method

A detailed discussion of calorimetric methods is outside the scope of this book. However, a further calorimetric method, designed originally by Tian and developed by Calvet [31], is also based on the differential principle. In this method the major part of the heat liberated in the calorimeter is removed by thermocouples to the external block. The sensor is a system of thermocouples connected in series, the junctions of which are alternately connected to the container wall, the sample, and the external block. The junctions are separated from the walls by a thin layer of dielectric. This detector of the heat flow should remove the heat from the container in the direction of the block [or in the opposite direction, depending on the temperature difference which exists at a given moment between the internal (i.e. container) and external (i.e. block) junctions] and also supply an e.m.f. proportional to the temperature difference, and thus also to the heat flow. As the thermocouple leads have a low heat capacity the values of the heat flow and of e.m.f. are interrelated and the e.m.f. at each moment is proportional to the heat transfer between the vessel and the block. The differential connection of two identical thermo-couple systems located symmetrically eliminates parasitic e.m.f.'s which may occur as a result of the fluctuating block temperature and which might distort the zero line. In practice, each detector of the heat flow is constructed from two thermocouples interconnected and having an unequal number of contact sites. This construction permits different connections of sensors to the recorder, allowing the sensitivity to be changed, as well as a connection between that thermocouple which is not connected to the recorder, and the direct current source. The Peltier effect (i.e. the liberation of heat from the thermocouple junction when a current is passed) may be used in two ways, either for the creation of a temperature difference between the container and the block, allowing the measurement of the specific heat and of thermal conductivity, or for the compensation of internal thermal processes in the measuring vessel (either partially or completely). Hence, only the heat of reaction itself can be followed. The calorimeter is calibrated by the introduction of a resistance into the measuring vessel. The arrangement of the thermocouples, which is fundamental to this method, and the parameters of commercially produced Calvet calorimeters, are given in Chapter 5.

3.9.8. Special methods of DTA and related calorimetric methods

From the quantitative point of view differential thermal analysis offers great possibilities in a number of special applications in various branches of science. Metallurgy at least should be mentioned briefly, where quantitative DTA has been used for the measurement of enthalpy changes during the solidification and melting of various metals and alloys. Neumann and Predel [113] developed a method for the determination of the heat of solidification of metals and alloys, based on quantitative DTA. The measuring cell, comprising an insulated tube containing the analysed material and the detecting thermocouple, and the internal heater for calibration are located in an aluminium block containing the reference thermocouple. The aluminium block is placed in a well-insulated muffle furnace. The cooling curves which show the effect of solidification of the alloy analysed allow the determination of the enthalpy changes in pure components, as well as in liquid mixtures. Calibration was done electrically by means of a heating coil located near the thermocouple in the sample. The method was checked with pure components and gallium and cadmium alloys.

Lugscheider [96] developed a method of quantitative DTA for the determination of heats of fusion of highly reactive alloys. He discussed the possible errors of this method and deduced a relationship for quantitative measurements, based on the effect of the heat transfer by radiation at high temperatures, which he checked by calibration measurement. For the measurement he used three carrier thermocouples on which round-bottomed tantalum crucibles were placed. All three crucibles were pressed on the junctions of the thermocouples by the same force by means of a plate and a spring, thus securing equal contact. According to his findings the calculated enthalpies vary with T^4 (in view of the effect of radiation). Hence, the determination of heats of reaction by means of DTA seems useful only up to 1000 °C. The results obtained at higher temperatures should be considered only as approximate.

As DTA has wide use in a number of fields, various special methods are met in the literature, which usually have a limited application. Only a few will be mentioned here, e.g. samples have been used in the form of pellets or briquettes, which gave rise to a better dehydration effect and an increase in the resolving power [112, 155]. DTA at increased pressure, as described by Berg and Rassonskaya [17], is a very specialized technique, as are methods using closed vessels. They may contribute to the elucidation of complex reactions. Elimination of the reference sample and the location of the differential thermocouple junction in the protecting tube closely under the analysed sample, as is done by some workers [62, 112], may be advantageous in the case of some combined methods (DTA + TG).

Increased attention should be given to combined methods capable of carrying out several types of analyses in one experiment. Recently such methods have been widely used and they are increasingly met in commercial apparatus, demonstrating the success of the method. The aim of these methods is to obtain a larger amount of information with the elimination of some of the discrepancies between the results obtained by the same methods when used separately. The combination of DTA and TG has become most popular. It was developed by a number of authors [21, 22, 67, 102, 119, 121] and commercial apparatus is readily available, e.g. the Derivatograph and apparatus made by various firms in West Germany, Great Britain, U.S.A. and Japan. With some apparatus the DTG curve and the analysis of evolved gases are also recorded. Pai Verneker and Maycock [117] developed simultaneous TG and DTA methods of analysis of explosives. DTA is also often combined with mass spectrometry [75, 76]. Visual observation of the sample during DTA may considerably simplify the interpretation of the results in cases where a change in physical state or in colour occurs.

Although various combined methods (for example TG + DTA) are widely used and have considerable advantages with respect to time saving and the improvement of the information obtained, their use should be based on a thorough knowledge of the fundamentals and problems of both methods. As has been explained in the preceding sections the results are dependent on a series of experimental factors, such as rate of heating, geometry, amount of sample, furnace atmosphere, etc., which should be carefully controlled in the case of a combined analysis if the results are to be interpreted correctly. However, the combination of both methods (TG +

DTA) is advantageous in analytical applications as it affords two types of information obtained under identical experimental conditions, though in the case of other applications this combination may even have certain disadvantages. It should be borne in mind that a combination of methods leads to a compromise in the optimum experimental conditions for each.

In conclusion, it should be noted that combination of the methods of thermal analysis is not limited to the two methods mentioned, i.e. DTA and TG, but that in special cases these fundamental methods may be combined with other methods. Thus, DTA has been combined with evolved gas analysis in the case of the analyses of special steels [4, 7], Paulik [120] has combined DTA, TG and DTG measurements with dilatometry, Chiu (60) has developed a method combining TG, DTG and DTA with electrothermal analysis and Balek has combined DTA with the emanation method [179, 180]. The list of these methods could be continued, which shows the wide application of present-day apparatus in the field of thermal analysis.

Among special methods of DTA the one using low temperatures (from liquid helium, or more usually liquid nitrogen, temperature to about $+20\ °C$) is increasingly more used, mainly in the chemistry of polymers, and its spread is facilitated by the accessibility of commercial apparatus. The following references represent only a short enumeration of papers in this field [181 − 190].

References

1. AGARWALA R. P., NAIK M. C. *Anal. Chim. Acta* **24**, 128 (1960).
2. ALLISON E. B. *Clay Minerals Bull.* **2**, 242 (1955).
3. ARENS P. L. *A Study of the Differential Thermal Analysis of Clays and Clay Minerals,* Excelsior Foto Offset, The Hague (1951).
4. AYRES W. M., BENS E. M. *Anal. Chem.* **33**, 568 (1961).
5. BAYLISS P., WARNE S. ST. *Am. Mineralogist* **47**, 775 (1962).
6. BANDO Y., KIAYAMA M., TAKADA T., KACHI S. *Japan J. Appl. Phys.* **4**, 240 (1965).
7. BANDI W. R., STRAUB W. A., BUYOK E. G., MELNICK L. M. *Anal. Chem.* **38**, 1336 (1966).
8. BÁRTA R., and CO-AUTHORS *Příručka pro cvičení v laboratoři technologie silikátů.* Institute of Chemical Technology, Prague, 1954.
9. BÁRTA R. *Silikáty* **13**, 185 (1969).
10. BARRALL E. M., ROGERS L. B. *Anal. Chem.* **34**, 1101 (1962).
11. BARRALL E. M., ROGERS L. B. *Anal. Chem.* **34**, 1106 (1962).
12. BARRALL E. M., PORTER R. S., JOHNSON J. F. *Anal. Chem.* **37**, 1053 (1965).
13. BARSHAD I. *Am. Mineralogist* **37**, 667 (1952).
14. BEHAR M. F. *Manual of Instrumentation.* Instruments Publishing Co. Pittsburg, 1932, Vol. 2.
15. BERG L. G. *Dokl. Akad. Nauk SSSR* **49**, 648 (1945).
15a. BERG L. G. *Proc. III Anal-Chem. Conf. Budapest,* 1970, Vol. 2, p. 211.
16. BERKELHAMER L. H. *U.S. Bur. Mines, Rept. Invest.* No. 3763 (1944).
17. BERG L. G., RASSONSKAYA I. S. *Dokl. Akad. Nauk SSSR* **81**, 855 (1951).
18. BLAŽEK A., CÍSAŘ V., ČÁSLAVSKÁ V., ČÁSLAVSKÝ J. *Collection Czech. Chem. Commun.* **25**, 2419 (1960).
19. BLAŽEK A., HALOUSEK J. *Hutnické Listy* **14**, 244 (1959).
20. BLAŽEK A., DOUSEK J. *DTA elektrod palivových článků.* Polarographic Institute, Czechoslov. Acad. Sci. Prague. Unpublished.
21. BLAŽEK A., CÍSAŘ V. *Silikáty* **4**, 52 (1960).
22. BLAŽEK A. *Hutnické Listy* **12**, 1096 (1957).
23. BLUMBERG A. A. *J. Phys. Chem.* **63**, 1129 (1959).
24. BOHON R. L. *Anal. Chem.* **35**, 1845 (1963).
25. BOERSMA S. L. *J. Am. Ceram. Soc.* **38**, 281 (1955).
26. BOUFFANT L. *Bull. Soc. Franc. Céram.* **71**, 65 (1966).
27. BORCHARDT H. J., DANIELS F. *J. Am. Chem. Soc.* **79**, 41 (1957); *J. Inorg. Nucl. Chem.* **12**, 252 (1960).
28. BRAMAO L., CADY J. G., HENDRIKS S. B., SWERDLOW M. *Soil Sci.* **73**, 273 (1952).
29. DE BRUIN C. M. A., VAN DER MAREL H. W. *Geol. en Mijnbouw* (N. S.) **16**, 69 (1954); **16**, 407 (1954).
30. BURGESS G. K. *Bull. Natl. Bur. Std.* **5**, 199 (1907—1909).

31. CALVET F., PRAT H. *Microcalorimetrie. Applications phys. chim. et biologiques.* Masson, Paris, 1956.
32. CAMPBELL C., GORDON S., SMITH C. L. *Anal. Chem.* **31,** 1188 (1959).
33. CARPENTER H. C. H., KEELING B. F. E. *J. Iron Steel Inst. (London)* **65,** 224 (1964).
34. CARTHER A. R. *Am. Mineralogist* **40,** 107 (1955).
35. CUNNINGHAM R. C., WELD H. M., CAMPBELL W. P. *J. Sci. Instr.* **29,** 252 (1952).
36. DAVID D. J. *Anal. Chem.* **36,** 2162 (1964).
37. DEAN L. A. *Soil Sci.* **63,** 95 (1947).
38. DEMPSTER P. B., RITCHIE P. D. *Nature* **169,** 538 (1952).
39. DUSOLLIER G., ROBREDO J. *Verres Refractaires* **21,** 550 (1967).
40. ELLIS B. G., MORTLAND M. M. *Am. Mineralogist* **47,** 371 (1962).
41. ERIKSSON E. *Kgl. Lantbruks Högskol Ann.* **19,** 127 (1953); **20,** 117 (1954); **21,** 189 (1954).
42. FOLVARDI-VOGL M., KLIBURSKY B. *Acta Geol. Acad. Sci. Hung.* **5,** 187 (1958).
43. FOLTZ C. R., McKINNEY P. V. *Anal. Chem.* **41,** 687 (1969).
44. FREDERICKSON A. F. *Am. Mineralogist* **39,** 1023 (1954).
45. FREEMAN E. S., EDELMAN D. *Anal. Chem.* **31,** 624 (1959).
46. FREEMAN E. S., CARROLL B. *J. Phys. Chem.* **62,** 394 (1958).
47. GARN P. D. *Thermoanalytical Methods of Investigation,* Academic Press, New York, London, 1965. p. 223—265.
48. GARN P. D., FLASCHEN S. S. *Anal. Chem.* **29,** 271 (1957).
49. GARN P. D. *Anal. Chem.* **37,** 77 (1965).
50. GARN P. D. *Thermoanalytical Methods of Investigation* [47] p. 171.
51. GARN P. D. *Proc. Intern. Symposium Microchem. Techniques,* Pennsylvania State College 1961. Wiley, New York 1962.
52. GÉRARD-HIRNE J., LAMY C. *Bull. Soc. Franc. Ceram.* 26 (1951).
53. GRIM R. E. *Ann. N. Y. Acad. Sci.* **53,** 31 (1951).
54. GRIMSHAW R. W., HEATON E., ROBERTS A. L. *Trans. Brit. Céram. Soc.* **44,** 76 (1945).
55. HAUSER R. E. *Dissertation,* N.Y. College of Ceramics, Alfred, New York 1953.
56. HENDRICKS S. B., ALEXANDER L. T. *Soil Sci.* **48,** 257 (1939).
57. HOGAN V. D., GORDON S. *Anal. Chem.* **32,** 573 (1960).
58. CHIU J. *Anal. Chem.* **35,** 933 (1963).
59. YAGFAROV M. S. *Russ. J. Inorg. Chem.* (English Transl.) **6,** 1236 (1961).
60. CHIU J. *Anal. Chem.* **39,** 861 (1967); *Anal. Chem.* **35,** 933 (1963).
61. JANKOWSKI B. *Silikattechnik* **17,** 20 (1966).
62. JONCIN M. J. BAILEY D. R. *Anal. Chem.* **32,** 1578 (1960).
63. DE JONG G. J. *J. Am. Ceram. Soc.* **40,** 42 (1957).
64. KERR F. P., KULP J. L. *Am. Mineralogist* **33,** 387 (1948).
65. DE KEYSER W. L. *Ann. Mines. Belg.* **39,** 985 (1939); **40,** 357 (1940); **40,** 711 (1940).
66. KING JR. H., CAMILI C. T., FINDEIS A. F. *Anal. Chem.* **40,** 1330 (1968).
67. KISSINGER H. E., McMURDIE H. F., SIMPSON B. S. *J. Am. Ceram. Soc.* **39,** 168 (1956).
68. KISSINGER H. E. *J. Res. Natl. Bur. Std.* **57,** 217 (1956).
69. KISSINGER H. E. *Anal. Chem.* **29,** 1702 (1957).
70. KRACEK F. C. *J. Phys. Chem.* **34,** 225 (1930).
71. KRAWETZ A. A., TOVROG T. *Rev. Sci. Instr.* **33,** 1465 (1962).
72. KRAIS ST. *Silikáty* **7,** 52 (1963).
73. KRONIG R., SNOODIJK F. *Appl. Sci. Research* A3, 27 (1951).
74. KURNAKOV N. S. *Zh. Fiz. Khim.* **36,** 841 (1964).

75. LANGER H. G., GOHLKE R. S., SMITH D. H. *Anal. Chem.* **37**, 433 (1965).
76. LANGER H. G., GOHLKE R. S. *Anal. Chem.* **35**, 1301 (1963).
77. MCLAUGHLIN R. J. W. *Trans. Brit. Ceram. Soc.* **60**, 177 (1961).
78. LEHMANN H. *Ber. Deutsch. Keram. Ges.* **32**, 172 (1955).
79. LEHMANN H., DAS S. S., PAETSCH H. H. *Tonindustr. Ztg.* (Suppl. 1), 55 (1954).
80. LE CHATELIER H. *Compt. Rend.* **102**, 1243 (1886).
81. LE CHATELIER H. *Bull. Soc. Franc. Minéral. Crist.* **10**, 204 (1887).
82. LE CHATELIER H. *Compt. Rend.* **104**, 1443 (1887).
83. LE CHATELIER H. *Compt. Rend.* **104**, 1517 (1887).
84. LE CHATELIER H. *Z. Phys. Chem.* **1**, 396 (1887).
85. LE CHATELIER H. *Rev. Mét.* **1**, 134 (1904).
86. LODDING W., HAMMELL L. *Anal. Chem.* **32**, 657 (1960).
87. LUKASZEWSKI G. M. *Lab. Pract.* **14**, 1277, 1399 (1965).
88. LUKASZEWSKI G. M. *Lab. Pract.* **15**, 82 (1966).
89. LUKASZEWSKI G. M. *Lab. Pract.* **15**, 187 (1966).
90. LUKASZEWSKI G. M. *Lab. Pract.* **15**, 302, 306 (1966).
91. LUKASZEWSKI G. M. *Lab. Pract.* **15**, 431 (1966).
92. LUKASZEWSKI G. M. *Lab. Pract.* **15**, 551, 558 (1966).
93. LUKASZEWSKI G. M. *Lab. Pract.* **15**, 664 (1966).
94. LUKASZEWSKI G. M. *Lab. Pract.* **15**, 762, 771 (1966).
95. LUKASZEWSKI G. M. *Lab. Pract.* **15**, 861, 869 (1966).
96. LUGSCHEIDER W. *Ber. Bunsen Ges. Phys. Chem.* **71**, 228 (1967).
97. MACKENZIE R. C. *The Differential Thermal Investigation of Clays.* Mineralogical Society. London 1957.
98. MACKENZIE R. C. *Nature* **174**, 688 (1954).
99. MACKENZIE R. C. *Differential Thermal Analysis Data Index "Scifax".* Cleaver-Hume Press, London 1962.
99a. MACKENZIE R. C. *editor, Differential Thermal Analysis*, Vols. 1 and 2, Academic Press London 1970, 1972).
100. MAZIÈRES Ch. *Bull. Soc. Chim. France.* 1695 (1961).
101. MCADIE H. G. *Anal. Chem.* **39**, 543 (1967).
102. MCADIE H. G. *Anal. Chem.* **35**, 1840 (1963).
103. MERZHANOV A. G., DURAKOV N. I. *Zh. Fiz. Khim.* **40**, 811 (1966).
104. METTLER INSTRUMENTS A. G. *Documentation of Thermal Analyser Mettler,* Zürich 1968.
105. MURRAY P., WHITE J. *Trans. Brit. Cer. Soc.* **48**, 187 (1949); **54**, 137, 189, 204 (1955).
106. MURPHY C. B. *Anal. Chem.* **30**, 867 (1958).
107. MURPHY C. B. *Anal. Chem.* **32**, 168 R (1960).
108. MURPHY C. B. *Anal. Chem.* **34**, 298 R (1962).
109. MURPHY C. B. *Anal. Chem.* **36**, 374 R (1964).
110. MURPHY C. B. *Anal. Chem.* **38**, 443 R (1966).
111. MURPHY C. B. *Anal. Chem.* **40**, 380 R (1968).
112. NEWKIRK T. F. *J. Am. Ceram. Soc.* **41**, 409 (1958).
113. NEUMANN T., PREDEL B. *Electrochem.* **63**, 988 (1959).
114. NORTON F. H. *J. Am. Ceram. Soc.* **22**, 54 (1939).
115. O'NEILL M. J. *Anal. Chem.* **36**, 1238 (1964).
116. O'NEILL M. J. *Anal. Chem.* **38**, 1331 (1966).
117. PAI VERNEKER V. R., MAYCOCK J .N. *Anal. Chem.* **40**, 1325 (1968).
118. PAKULAK J. M., LEONARD G. W. *Anal. Chem.* **31**, 1037 (1959).

119. PAPAILHAU J. *Bull. Soc. Franc. Mineral. Crist.* **82,** 367 (1959).

120. PAULIK F. *Proc. Anal. Chem. Conf.* Vol. III, 333 (1966).

121. PAULIK F., PAULIK J., ERDEY L. *Z. Anal. Chem.* **160,** 241 (1958).

122. PILOYAN G. O., RYABCHIKOV I. D., NOVIKOVA O. S. *Nature* **212,** 1229 (1966).

123. PILOYAN G. O. *Vvedenie v teoryu termicheskogo analiza.* Izd. Akad. Nauk, Moscow 1964.

124. PLATO C., GLASGOW A. R. *Anal. Chem.* **41,** 330 (1969).

125. PROKS J. *Silikáty* **5,** 114 (1961).

126. REDFERN J. P., Editor, *Thermal Analysis Review,* Stanton Instruments, Ltd., London.

127. REICH L. *J. Appl. Polymer Sci.* **10,** 813 (1966).

128. ROBERTS-AUSTEN W. C. *Proc. Roy. Soc. (London)* A **49,** 347 (1891).

129. ROBERTS-AUSTEN W. C. *Proc. Inst. Mech. Engrs. (London)* **1,** 35 (1899).

130. ROGERS R. N. *Microchem. J.* **5,** 91 (1961).

131. ROWLAND R. A. *Calif. Dept. Nat. Resources, Div. Mines, Bull.* No. 169, 151 (1955).

132. ROWLAND R. A., LEVIS D. R. *Am. Mineralogist* **36,** 80 (1951).

133. ROWLAND R. A., JONAS E. C. *Am. Mineralogist* **34,** 550 (1949).

134. SALADIN E. *Iron and Steel Metallurgy and Metallography* **7,** 237 (1904).

135. SABATIER G. *Bull. Soc. Franc. Mineral. Crist.* **77,** 953, 1077 (1954).

136. SEWELL E. C. *Clay Minerals Bull.* **2,** 233 (1955).

137. SEWELL F. C. *Theory of DTA, Research Note, Building Research Station D.S.I.R.* London 1 (1952), II (1953).

138. SCHAFER G. M., RUSSELL M. B. *Soil Sci.* **53,** 353 (1942).

139. SCHWIETE H. E., ZIEGLER G. *Ber. Deut. Keram. Ges.* **35,** 193 (1958).

140. SMOTHERS W. J., YAO CHIANG M. S. *Handbook of Differential Thermal Analysis.* Chemical Publishing Comp., New York 1966.

141. SMYTH H. T. *J. Am. Ceram. Soc.* **34,** 221 (1951).

142. SPEIL S. *U.S. Bur. Mines, Rept. Invest.* No. 3764 (1944).

143. SPEIL S., BEUKELHAMMER L. H., PASK J. A., DAVIS B. *Bur. Mines Tech. Papers* 664 (1945).

144. SOULÉ J. L. *J. Phys. Radium* **13,** 516 (1952).

145. SCHULZE D. *Differentialthermoanalyse.* VEB Verlag der Wissenschaften, Berlin 1969, p. 87.

146. STONE R. L. *J. Am. Ceram. Soc.* **35,** 76 (1952); **35,** 90 (1952).

147. STONE R. L. *Proc. Second Natl. Conf. Clay Minerals* 315 (1953).

148. STONE R. L. *J. Am. Ceram. Soc.* **37,** 46 (1854).

149. STONE R. L. *Anal. Chem.* **32,** 1582 (1960).

150. SYUZO SEKI *Scientific Instrument News, Thermal Analysis Instruments.* Shimadzu Seisakusho Kyoto, 1969, Vol. 1, No. 1.

151. ŠATAVA V. *Silikáty* **5,** 68 (1961).

152. TAKAMURA T., KOEZUKA J. *Nature* **207,** 965 (1965).

153. VASSALLO D. A., HARDEN J. C. *Anal. Chem.* **34,** 132 (1962).

154. VOLD M. J. *Anal. Chem.* **21,** 683 (1949).

155. WALTON J. D. JR. *J. Am. Ceram. Soc.* **38,** 438 (1955).

156. WATSON E. S., O'NEILL M. T., JUSTIN J., BRENNER N. *Anal. Chem.* **36,** 1233 (1964).

157. WEAVER E. E., KEIM W. *Proc. Indian Acad. Sci.* **70,** 123 (1960).

158. WEBB T. L. *Nature* **174,** 686 (1954).

159. WENDLANDT W. W. *Thermal Methods of Analysis.* Interscience, New York, London, Sidney 1964.

160. WENDLANDT W. W. *in Technique of Inorganic Chemistry*, Vol. I., editors H. B. Jonassen and A. Weissberger, Interscience, New York 1963.
161. WENDLANDT W. W. *J. Chem. Educ.* **38**, 571 (1961).
162. WITTELS M. *Am. Mineralogist* **36**, 760 (1951).
163. WITTELS M. *Am. Mineralogist* **36**, 615 (1951).
164. YAMAMOTO A., YAMADA K., AKIAMA J., OKINO T. *J. Chem. Soc. Japan, Industr. Chem. Sect.* **71**, 2002 (1968).
165. YAMAMOTO A., AKIAMA J., OKINO T. *Japan Analyst* **17**, 1126 (1968).
166. ZAHRADNÍK L., ŠŤOVÍK M., ELIÁŠ M. *Chemické rozbory nerostných surovin* No. 12. *Diferenčně termická analyza.* Publ. House of the Czechoslov. Acad. Sci., Prague 1957.
167. ZÝKA J. *Analytická příručka*. SNTL, Prague 1967 (Chapter "Termická analyza").
168. BERGGREN G., ŠESTÁK J. *Chem. Listy* **80**, 561 (1970).
169. ŠESTÁK J., BERGGREN G. *Chem. Listy* **80**, 695 (1970).
170. SCHEDLING J. A. *Staub* **19**, 161 (1959).
171. MELLING R., WILBURN F. W., McINTOSH R. M. *Anal. Chem.* **41**, 1275 (1969).
172. DUSOLLIER G., ROBREDO J. *Verres Réfractaires* **21**, 550 (1967).
173. DUSOLLIER G., DÉROBERT M. *Verres Réfractaires* **23**, 10 (1969).
174. LEDRER E. *Linseis Journal* No. 2, 2–4 (1970).
175. TAYLOR L. J., WATSON S. W. *Anal. Chem.* **42**, 297 (1970).
176. REED R. L., WEBER B. S., GOTTFRIED B. S. *Ind. Eng. Chem. Fundamentals* **4**, 38 (1965).
177. ADAM G., MÜLLER F. H. *Kolloid Z., Z. Polymere* **192**, 29 (1963).
178. ROSICKÝ J. *Silikáty* **12**, 295 (1968).
179. BALEK VL. *Bull. Chem. Soc. Belgrade* **34**, 43 (1970).
180. BALEK VL. *Anal. Chem.* **42**, 16A (1970).
181. PROKS I., SISKE V. *Chem. Zvesti* **15**, 309 (1961).
182. MAZIÈRES C. *Ann. Chim.* 575, (1961).
183. REISMAN A. *Anal. Chem.* **32**, 1566 (1960).
184. ANIKIN A. G., DUGACHEVA G. M. *Doklady Akad. Nauk SSSR.* **135**, 634 (1960).
185. KATO C. *Yogyo Kyokai Shi* 67, 243 (1959).
186. BARRALL E. M., GERNERT J. F., PORTER R. S., JOHNSON J. F. *Anal. Chem.* **35**, 1837 (1963).
187. REDFERN G. P., TREHERNE B. L. *3rd ICTA Conference on Thermal Analysis*, Davos 1971, Abstracts of Papers I-5.
188. MAY J. C. H. *Thermal Analysis*, Davos 1971, Abstracts of Papers 1-7.
189. MILLER G. W. *Thermal Analysis* (editors R. F. Schwenker Jr. and P. D. Garn), Academic Press, New York, London, Vol. 1. 1969 p. 435.
190. MOORE R. *Thermal Analysis* (editors R. F. Schwenker Jr. and P. D. Garn). Academic Press, New York, London, Vol. 1. 1969 p. 615.

Chapter 4.

APPLICATIONS OF THERMAL ANALYSIS

4.1. TG AND DTA

In the preceding chapters the principles of the two basic methods of thermal analysis, TG and DTA, and procedures of obtaining meaningful results were outlined, together with a discussion of the conditions and factors affecting these results. Despite all their limitations and possible sources of errors, these techniques have in the past twenty years become valuable methods of investigation, and have found wide use in all fields of inorganic and organic chemistry, metallurgy, mineralogy, and many other areas. Applications of the methods of thermal analysis are listed below.

A. In the field of analytical chemistry:
  - (*a*) automatic gravimetry (TG)
  - (*b*) investigation of new compounds suitable for gravimetry (TG)
  - (*c*) investigation of thermal behaviour of compounds in various gaseous media (TG and DTA)
  - (*d*) following of the purity and the stability of compounds, by comparison with standards (TG and DTA)
  - (*e*) systematic study of the properties of prepared compounds (TG and DTA)
  - (*f*) study of dehydration and decomposition reactions (TG and DTA)
  - (*g*) study of new separation methods (TG and DTA)
  - (*h*) analysis of solid substances, especially minerals (TG and DTA)
  - (*i*) investigation of dissolution (TG and DTA)

B. In the field of inorganic chemistry (minerals, ceramics, metals, alloys, sintered carbides, glass, catalysts, construction materials, etc.):
  - (*a*) systematic study of prepared compounds (TG and DTA)
  - (*b*) study of chemical processes (TG and DTA)
  - (*c*) study of metallurgical reactions (TG and DTA)

(*d*) study of reaction kinetics (TG and DTA)
(*e*) study of stability and decomposition reactions (DTA and TG)
(*f*) study of heats of reaction and specific heats (DTA and DSC)
(*g*) study of corrosion reactions (TG and DTA)
(*h*) determination of phase diagrams (TG and DTA)
(*i*) study of purity (TG and DTA)
(*j*) study of crystallization and of the formation of phases (DTA)
(*k*) study of magnetic changes (Curie point) (TG and DTA)
(*l*) study of transformation changes (TG and DTA)
(*m*) study of the effect of radioactive irradiation (TG and DTA)

C. In the field of organic chemistry (drugs, plastics, explosives, detergents, dyes, textile fibres, fuels, celluloses, polymers, etc.):
(*a*) study of stability (TG and DTA)
(*b*) study of hydration and dehydration (TG and DTA)
(*c*) study of reaction kinetics (TG and DTA)
(*d*) study of polymerization reactions (DTA and DSC)
(*e*) study of melting and sublimation (TG and DTA)
(*f*) study of reaction and specific heats (DTA, DSC)
(*g*) study of catalytic activity (DTA)
(*h*) study of purity (DTA and TG)
(*i*) identification (TG and DTA)
(*j*) study of polymerization reactions (TG and DTA)
(*k*) study of oxidation stability (TG and DTA)
(*l*) study of dissolution (TG and DTA)

The list above, though not exhaustive, shows the wide scope of thermal analysis. The development of apparatus and experimental technique over the past twenty years has led to the publication of a great number of papers on this subject. A series of monographs on thermal analysis also shows the wide application of these methods. In analytical chemistry the book by Duval [34] is of prime importance, while in other fields the following are most valuable: Mackenzie's book [102] for clay minerals, Slade's [143] for organic polymers and Robredo's [131] for glasses. This list could easily be extended. Several review articles have also appeared on the application of these methods in various fields [163, 164], in addition to those mentioned in the introductory part of this book.

Instead of discussing here the applications of the methods in various fields the author considers it more useful to tabulate the data on the thermal behaviour of some important substances used mainly in calibration of apparatus and in analysis. The materials most commonly used have been selected, i.e. compounds suitable for gravimetry, or as temperature standards in thermogravimetry, for temperature calibration, for the calibration of the heat of reaction in DTA, and some mineral substances.

4.2. SELECTED EXAMPLES OF APPLICATIONS

4.2.1. Substances suitable for automatic gravimetry

"Automatic gravimetric analysis", as proposed by Duval, consists in using the thermogravimetric curve mainly for the determination of inorganic ions.

From the TG curve which has a horizontal part, suitability of a well-defined derivative (precipitate) for quantitative application can be evaluated. This method is of great importance in the analysis of mixtures where after precipitation and collection in a suitable crucible, a product may be subjected to TG, and various horizontal parts of the TG curve, corresponding to single compounds, may be used without the time-consuming separation used in classical gravimetry. However, only some precipitation reagents will be suitable for this method. Duval prefers organic reagents that yield complex compounds which are easily dried and do not require the heating at elevated temperatures which is necessary for ignition to oxides to achieve a form suitable for this method of determination. The sample (100−500 mg) should be heated in the form of a thin layer when only slightly moist. Duval [34] lists the precipitates which he considers are suitable for this determination.

Reference is often made to TG curves, for choosing suitable *static* heating conditions for drying or ignition of compounds. It is most important to check the validity of the selected temperature ranges experimentally, especially for work on the micro scale. A barely detectable change on a TG curve may cause a large error when isothermal conditions are used over longer periods of time. A classic case is dehydration of heteropoly acids. Duval [34] quotes temperatures of up to 840 ° as the upper limit for silicomolybdic anhydride, but on isothermal heating MoO_3 sublimes out of the precipitate at temperatures above 600° [113b]. Furthermore the weighing forms recommended by Duval are guaranteed suitable only if weighed on the thermobalance at the temperature indicated. If the ignited substance is to be cooled and weighed in air at room temperature, it must be tested for reversibility of dehydration (cf. Miller on calcium oxalate monohydrate [113a]).

4.2.2. Substances suitable for standardization in thermogravimetry

In Chapter 2 it was explained that in thermogravimetry the commonly employed expression "decomposition temperature" had no physical meaning. This term is usually used for the lowest temperature at which an observable weight change begins. This temperature depends, as has already been shown, on the parameters of the apparatus used, and it does not give an absolute value of the temperature below which the reaction rate is zero. For these reasons the concept of the "procedural decomposition temperature", discussed earlier, was introduced. Decomposition temperatures defined in this manner are usually employed for the definition of the thermal behaviour of substances suitable for the standardization of thermogravimetric equipment. A systematic study of such substances was carried out by Keattch [80], a member of the standardization committee of ICTA. Substances suitable for such standardization should have sufficient stability at room temperature, and good reproducibility of their thermal behaviour as characterized by the decomposition temperatures, and also sufficient reproducibility when samples of different origins are compared.

Studies of the use of standard substances in thermogravimetry are mainly directed towards eliminating the effects of using different apparatus, and obtaining comparable results. However, it is evident that as the temperature of the beginning of the decomposition on the TG curve is dependent, in addition to other effects, on the sensitivity of the apparatus, the use of these standards is recommended for the calibration itself. This is also recommended in cases where discrepancies in characteristic temperatures are observed. Moreover, the given apparatus can be calibrated on the basis of known, precisely defined reactions, not only according to the characteristic points on the temperature axis, but also with regard to the weight-change axis. Some of the recommended TG standards are listed in Table 4.1.

4.2.3. Substances suitable for the calibration of temperature and heats of reaction in DTA

As shown in Sections 3.5 and 3.7.2, correct interpretation of DTA results depends on the evaluation of the temperature axis and the area under the curve. For this purpose substances are used which have a defined temperature of transformation. For the calibration of the peak area, which is proportional to the enthalpy change of the observed transformation, substances are used which undergo changes which have a well-defined enthalpy change. It is important that the calibration substance used should have its peak in nearly the same region as the investigated peak. This means that a choice of standards should be available which would cover the maximum temperature range and would also provide a choice of reactions with weak and strong heat effects. As transformations of equal enthalpy change occurring at different places on the temperature axis display effects of different amplitude,

Table 4.1

Selected Compounds Suitable for Calibration of Thermobalances [80]

Compound	Characteristic decomposition temperature °C
Aluminium ammonium sulphate 12-hydrate	50
Barium hydroxide 8-hydrate	50
Oxalic acid dihydrate	50
Zinc acetate dihydrate	50
Zinc sulphate heptahydrate	50
Aluminium potassium sulphate 12-hydrate	50
Copper(II) sulphate pentahydrate	52
Cadmium acetate dihydrate	60
Magnesium acetate tetrahydrate	60
Copper(II) sulphate pentahydrate	86
Potassium oxalate dihydrate	80
Potassium oxalate monohydrate	90
Boric acid	100
Oxalic acid	118
Copper (II) acetate monohydrate	120
Ammonium dihydrogen phosphate	185
Tartaric acid	180
Sucrose	205
Copper(II) sulphate pentahydrate	235
Zinc sulphate heptahydrate	250
Sodium hydrogen tartrate monohydrate	240
Potassium hydrogen tartrate	260
Potassium hydrogen phthalate	245
Cadmium acetate dihydrate	250
Magnesium acetate tetrahydrate	320
Potassium hydrogen phthalate	370
Barium acetate	445
Sodium hydrogen tartrate monohydrate	545
Potassium hydrogen phthalate	565
Copper(II) acetate monohydrate	1055
Copper(II) sulphate pentahydrate	1055

the calibration should be carried out in the temperature range where the substance under investigation undergoes transformation. For this reason substances of which the heats of fusion, decomposition, dehydration or transformation cover the temperature interval between 20 and 1000 °C were chosen for calibration purposes. It should be noted that decomposition and dehydration reactions are the most dependent on experimental conditions and are, therefore, less suitable for calibration purposes.

During calibration the standard may be placed at the level of the thermo-couple junction in the form of a thin layer sandwiched between layers of inert material, or it may be homogenized with the latter. However, the best method is to use the standard substance as a reference which gives on the DTA curve a temperature effect of the opposite sign. In special cases, for quantitative evaluation it is possible to mix the calibration substances with the inert reference sample in various concentrations, until compensation of the effects is achieved, which directly permits a single quantitative determination of ΔH. In addition to the calibration of the temperature axis on the basis of the known value of T_0 or T_m of the calibration substance, it is possible to calibrate the apparatus by a simple thermochemical calculation if the sample weight and the heat of transformation are known:

$$B = \int_{t_2}^{t_3} \Delta T \, dt = \Delta Q . A \qquad (3.32)$$

In this equation the proportionality constant, A, is dependent on the arrangement of the apparatus, the sensitivity and the temperature. Although the value of A increases with temperature (see Fig. 4.1) which is due to the increasing amount of heat transfer by radiation, for the calculation of B the value of A may be taken as constant within the narrow temperature range

Table 4.2

Substances Suitable for Calibration of Temperature and Heat of Reaction in DTA [168, 2]

Substance†	Type of change*	Temperature of maximum T_m, °C	Heat of change cal/g
Nitrobenzene	M	5.7	22.5
Gallium	M	29.8	1335 †
Ammonium nitrate	T	32.1	4.75
	T	85	
	T	125.2	16.6
	M	169.6	16.2
p-Nitrotoluene	M	51.5	
Palmitic acid	M	63.0	51
Stearic acid	M	64.0	47.6
Naphthalene	M	80.2	
m-Dinitrobenzene	M	90.0	24.7
o-Dinitrobenzene	M	117	32.3
Barium chloride dihydrate	Dh	(120)	119
Strontium chloride dihydrate	Dh	(120)	311
Sodium thiosulphate pentahydrate	Dh	(120)	269

Substance†	Type of* change	Temperature of maximum T_m, °C	Heat of change cal/g
Benzoic acid	M	122.2	33.9
Potassium nitrate	T	127.7	13.9
	M	337	25.3
Calcium nitrate tetrahydrate	Dh	(130)	230
Magnesium sulphate heptahydrate	Dh	(130)	402
Gypsum	Dh	(140)	160
Silver iodide	T	147	
Adipic acid	M	151.4	
Indium	M	157	
Silver nitrate	T	160	
	M	212	
	Dc	400	210
Silver sulphide	T	177	950†
Anisic acid	M	182.9	
2-Chloroanthraquinone	M	209	
Carbazole	M	245.3	
Lithium nitrate	M	254	88.4
			2650†
Bismuth	M	271	
Sodium nitrate	T	273	9.5
	M	310	45.3
	Dc	(667)	776
Anthraquinone	M	284.0	
Potassium perchlorate	Dc	299.5	26.2
Lead	M	327.4	5.89
Potassium dichromate	M	398	
Cadmium carbonate	Dc	400	49.3
Zinc carbonate	Dc	425	114
Silver sulphate	T	432	6.1
	M	652	
Silver chloride	M	436	22.0
Lithium sulphate	T	560	
	M	810	
Quartz	T	573	4.83
Potassium sulphate	T	583	11.3
	M	1069	52.6
Sodium tungstate, anhydrous	T	580	
Sodium molybdate, anhydrous	T	642	
	M	687	
Manganese(II) chloride	M	650	71.4
Aluminium	M	658.8	95.2
			2570†
Sodium nitrate	Dc	(667)	776
Sodium chloride	M	804	117
Barium carbonate; witherite	Dc	810	23.3
Silver	M	960.8	25.2
			2728†
Potassium sulphate	M	1069	52.6

* Abbreviations:
T — transformation, M — melting, Dc — decomposition, Dh — dehydration

† The hydration and decomposition temperatures of carbonates are dependent on experimental conditions and therefore unsuitable for temperature calibration. In the table they are put in brackets. Heat of reactions indicated by a dagger are expressed in cal/mole.

of the transformation investigated. Calibration substances (e.g. KNO_3) which give several characteristic points and thus also several calibration values by means of which the temperature dependence of the proportionality constant A may be determined, are therefore the most useful. The main condition for the use of calibration substances directly as reference substances is that both analysed and reference substances should have comparable heat capacities and thermal conductivities. This condition may be achieved by diluting the calibration substance with a suitable inert material.

Calibration materials should have sufficient stability at room temperature, good reproducibility of their thermal behaviour, and reproducibility among samples of various origin. In Table 4.2 some substances are listed which have been recommended at various times for calibration purposes in DTA. More recently the standardization committee of ICTA have considered this problem [101a] and have recommended eight compounds (Table 4.3) as provisional DTA temperature standards. These compounds are now commercially available in a guaranteed state of purity and may be obtained from the National Bureau of Standards as a complete set covering the temperature range from 120 or 250 to 925 °C or as subsets of four, covering the low, medium and high temperature ranges. The ICTA standardization committee has also dealt with the standardization of nomenclature [109a, 101c], and its most recent recommendations have been on apparatus

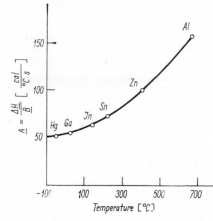

Fig. 4.1 Calorimetric calibration curve for DTA apparatus. (According to the advertising leaflet of DuPONT 800 Differential Thermal Analyser, by permission)

Table 4.3.

ICTA Standardization Committee Provisional DTA Temperature Standards [101a, 101b, 101c]

Standard	Transition temperature, °C
KNO_3	127.7
$KClO_4$	299.5
Ag_2SO_4	412*
SiO_2	573
K_2SO_4	583
K_2CrO_4	665
$BaCO_3$	810
$SrCO_3$	925

* Value doubtful; considerable evidence for 430 ± 2 °C

and curve nomenclature. Both its reports on recommended standard nomenclature are reproduced in Appendix II. That is the reference source of this list. In addition to substances listed there, an additional 100 organic substances, especially for the range of lower temperatures (20−235 °C), can be found in the paper by Plato and co-workers [156]; the heats of melting and the purity were determined by the DSC method. These substances include amides, carbamates and heterocyclic compounds. In Table 4.4 heats of decomposition and dehydration reactions of certain substances, determined by DTA by the calibration method, are given.

Table 4.4

Heats of Reaction of Some Decomposition and Dehydration Reactions used for Standardization in DTA [2]

Substance	Decomposition temperature on DTA curve, °C	Heat of decomposition cal/g
Calcite	700—830	465
Silver nitrate	370—470	212
Sodium nitrate	600—720	770
Brucite	365—455	332
Gibbsite	258—360	259
Goethite	314—396	105
Kaolinite	455—642	253
Halloysite	430—550	166
Ca-Montmorillonite	557—723	67
	816—908	26
Mg-Illite	400—695	64
	790—950	15
Gypsum	140*	164†
Calcium nitrate tetrahydrate	130*	234†
Barium chloride dihydrate	120*	120†
Magnesium sulphate heptahydrate	130*	395†
Strontium chloride hexahydrate	120*	300†
Sodium thiosulphate pentahydrate	120*	260†

* Temperature of desorption of water
† Heat of desorption of water

4.2.4. Thermal behaviour of some important mineral substances

In view of the fact that thermal analysis, mainly DTA, has found wide application in the analysis and the identification of substances in mineralogy, the thermal behaviour of some technically important mineral substances is given in Table 4.5. Attention has been given to carbonates, oxides, sulphides, sulphates, halides, nitrates, phosphates and silicates. The characteristic temperatures of the peaks (T_m) are given and the starting points (T_0) are also given where it was possible to find them in the literature.

In concluding this chapter on the application of the methods of thermal analysis, some other areas in which these methods are used will be mentioned briefly. In metallurgy DTA has been used for quantitative evaluation of heat changes in metallurgical reactions, reaction mechanisms have been followed on model systems, and in the case of pure materials transformation changes and enthalpy changes during melting have been studied. Numerous investigations have been made on ores, slags and metals under reducing and oxidizing conditions [13a, 16a, 30a, 72a 124a, 129a, 143a]. Thermite processes have also been investigated by means of DTA, as well as phase equilibria and phase diagrams. [20a, 28a, 47a, 130a, 141a, 142a, 147b, 156a, 160a, 161a].

Silicate chemistry, including construction materials, ceramics, and glasses, represents a field in which thermal analysis is widely used. Its use for the identification of mineral substances and the analysis of mixtures is evident from Table 4.5. In the case of glasses considerable work has been done on the thermal behaviour of the starting materials of glass production and on the thermal behaviour of glass itself which is characterized by a series of typical intervals within which its physical, chemical and structural properties change [131]. DTA is also used in the analysis of cements [99a, 131a, 153a, 166] for the investigation of raw materials [6, 15, 16, 139a], the production process, [86e, 124a] the formation of calcium and aluminium silicates, and the influence of additives on their formation [19b, 69a, 69b]. Dehydration can also be studied by this method [19a, 100a, 140a].

**The next 42 pages
tabulate the thermal behaviour
of many of the mineral substances
of importance in industry and technology**

Table 4.5.

Thermal behaviour of some technically

Substance	Temperature of the peak T_m °C	Temperature of the break-point T_0 °C	Effect	Reaction
BORATES				
B(OH)$_3$ Sassolite	157	85	endo large	volatilization
Na$_2$B$_4$O$_7$. 10 H$_2$O Borax	73 82 137	50	endo large	turbulent dehydration
Ca$_2$B$_6$O$_{11}$. 5 H$_2$O Inyoite	400	338	endo large	turbulent dehydration

NITRATES

In view of their appreciable solubility their presence is usually unimportant

KNO$_3$ Nitrokalite (Nitre)	127	117	endo large sharp	phase transition orthorhombic trigonal
NaNO$_2$ Nitronatrite (Nitratine)	315	300	endo small sharp	transformation

PHOSPHATES

Their composition is very variable and therefore the DTA curves are poorly reproducible

Al$_6$(PO$_4$)$_4$ (OH)$_6$. 9 H$_2$O Wavellite	375 315	150	endo large endo medium	dehydration
Fe$_3$(PO$_4$)$_2$. 8 H$_2$O Vivianite	260 330 380		endo large endo small endo small	− 5 H$_2$O − 2 H$_2$O − 1 H$_2$O

[1]) Over 480 °C thermocouples are attacked.

[2]) In the case of natural samples the melting effect is complicated by the impurities present. The substance is used for calibration in DTA.

important mineral substances

Side effect				Reference	Remarks
Temperature of the peak T_m, °C	Temperature of the break-point T_0 °C	Effect	Reaction		
				[102]	
				[102]	[1]
				[10], [102]	
324	307	endo large sharp	melting	[66]	
427	420	endo small shallow	melting	[2]	[2]
700		exo small	formation of Al, P tridymite	[70], [110]	[3]
	550	exo medium	oxidation of Fe^{2+}	[110]	[4]

[3] The position of the small exo effect is very unstable.
[4] Exothermic reactions at over 550 °C are connected with the oxidation of Fe^{2+} and the course depends on the pretreatment of the sample and on granulation.

Table 4.5

Substance	Main effect			
	Temperature of the peak T_m °C	Temperature of the break-point T_0 °C	Effect	Reaction
$AlPO_4 . 2 H_2O$ Variscite	240		endo medium	dehydration
$K_2Al_2(OH)_2(PO_4)_2$ Lazulite	220		endo medium	dehydration
$3 Al_2O_3 . P_2O_5 .$ $18 H_2O$ Evansite	190		endo large	dehydration

HALOGEN MINERALS

$NaCl$ Hallite		800	weak drift to the endo side	melting
KCl Sylvine		770	weak drift to the endo side	melting
Na_3AlF_6 Cryolite	565	528	endo large sharp sharp	transformation monoclinic cubic
$KMgCl_3 . 6 H_2O$ Carnallite	160 237	110 200	endo large endo large	dehydration dehydration
$Cu_2Cl(OH)_3$ Atacamite	280 380	250 310	endo small endo large	dehydration of $Cu(OH_2)$

[5]) The pure material gives a sharp inflection.

[6]) The pure material gives a sharp inflection.

[7]) Over 750 °C an endo effect connected with the corrosion of the thermocouple was observed.

(continued)

Side effect				Reference	Remarks
Temperature of the peak T_m, °C	Temperature of the break-point T_0 °C	Effect	Reaction		
				[10], [110]	
780		endo medium	formation of	[110]	
925		exo small	crystobalite		
350		endo small	dehydration	[110]	
525		endo small			
840		exo small	recrystallization		
				[2], [10], [121]	5)
				[10], [135]	6)
				[77]	7)
427	420	endo medium	dehydration	[132]	8)
437		endo medium			
480		endo medium			
606		endo medium	melting		
498–500	450	endo large	melting of $CuCl_2$	[43]	9)

8) Dehydration takes place stepwise, in the 100—280 °C region it gives two large effects, in the 420—500 °C it gives three small effects.

9) In the case of paratacamite, $Cu_2(OH)_3Cl$, the endo effect of dehydration occurs at a lower temperature (310 °C).

Table 4.5

OXIDES and HYDROXIDES

Oxides of Fe, Mn, and Al are often found with clay minerals, either as finely dispersed particles, solid solutions or surface films. Their frequent presence may distort the DTA curves of pure substances by their temperature effects. A knowledge of their thermal behaviour is therefore necessary when mineral substances are analysed.

When various polymorphous transformations of SiO_2 are observed, DTA is one of the most useful methods of thermal analysis. The transformation of quartz to both high-

Substance	Main effect			
	Temperature of the peak T_m °C	Temperature of the break-point T_0 °C	Effect	Reaction
HFeO$_2$ α-FeO . OH Goethite	300 to 400		endo medium	
β-Fe$_2$O$_3$. H$_2$O β-FeO . OH	125 250 280–300		endo medium $\}$ endo very small $\}$ endo very small $\}$	dehydration $\to \alpha$-Fe$_2$O$_3$ $+ \beta$-Fe$_2$O$_3$
γ-FeO . OH Lepidocrocite	300–350 max. 305	280	endo medium narrow	dehydration
γ-Fe$_2$O$_3$. H$_2$O	120	80	endo small	dehydration
α-Fe$_2$O$_3$ Hematite				
γ-Fe$_2$O$_3$ Maghemite				

[10]) The T_m value depends on the particle size and in the case of current samples it varies between 380 and 405 °C. In the case of very fine synthetic samples T_m is less than 300 °C. For well crystallized samples, T_m is about 390 °C. Natural limonites are composed of finely granulated goethite and have their endo effect at a somewhat lower temperature. The two-step reaction which appears sometimes is believed to be due to the formation of a layer of anhydrous oxide, inhibiting dehydration.

[11]) The curves of various samples often differ appreciably mainly with respect to the formation of exo effects, this is explained by the different content of Cl$^-$ ions. With well washed samples the exo effect is observed at 450 °C.

(continued)

temperature modifications, crystobalite and tridymite, is relatively slow. The reverse transformation to quartz is still slower and therefore experimentally demonstrable only with difficulty. In contrast to this all three polymorphous forms display during heating at relatively low temperatures a rapid and thermally poor reversible $\alpha \rightleftharpoons \beta$ transformation which is the basis of their qualitative and quantitative determination by means of DTA. Small energetic changes accompanying this transformation require a special method and very sensitive apparatus.

Side effect				Reference	Remarks
Temperature of the peak T_m, °C	Temperature of the break-point T_0 °C	Effect	Reaction		
				[41], [53], [77], [88], [128], [155]	[10]
450		endo medium sharp	→ α-Fe_2O_3	[103], [104], [159]	[11]
275		exo small sharp			[11]
508–570		exo small	$\gamma \rightarrow \alpha$-$Fe_2O_3$	[10], [88],	[12]
1444		endo small	melting of	[104], [49],	
–		reversible	Fe_2O_3	[103], [156]	
275	230	endo small		[10], [103]	[13]
320	300				
1360–1440		endo small	melting	[10]	[14]
–					
510—570		exo medium	$\gamma \rightarrow \alpha$-$Fe_2O_3$	[19]	[15]

[12]) The curves are more reproducible than in the preceding case. Sometimes two exo effects are observed, at 561 and 817 °C. The first one corresponds to $\gamma \rightarrow \alpha$ transformations and is catalysed by moisture, the second is not yet explained. Natural samples show a large scatter. The T_0 value depends on the particle size.

[13]) Only synthetic sample available. The curve is similar to that of natural hydrohematite.

[14]) In the 0–1000 °C temperature range the effect is not suitable for identification.

[15]) The curve has the same course as for lepidocrocite, except for dehydration. Only curves for synthetic material are available.

Table 4.5

Substance	Main effect			
	Temperature of the peak T_m °C	Temperature of the break-point T_0 °C	Effect	Reaction
Fe_3O_4 Magnetite	275–371	275	exo medium	$\rightarrow \gamma\text{-}Fe_2O_3$
FeO Wüstite	294		exo large sharp	$\rightarrow \alpha\text{-}Fe_2O_3$
$Fe_2O_3 \cdot n\,H_2O$ Amorphous gels formed by precipitation	250–500	200	exo sharp	crystallisation of amorphous to $\alpha\text{-}Fe_2O_3$ ($\beta\text{-}Fe_2O_3$)
$\alpha\text{-}Al(OH)_3$ Bayerite (according to the USA system β-trihydrate)	310–315	250–270	endo large	dehydration
$\gamma\text{-}Al(OH)_3$ Gibbsite (according to the USA system α-trihydrate)	320–330	270	endo small	dehydration to $\alpha\text{-}Al_2O_3$

[16]) As it is a spinel, Fe^{2+} and Fe^{3+} ions may be substituted by other isomorphous ions. Synthetic and natural samples differ appreciably. Well shaped curves are only obtained with synthetic samples. Natural samples give only a small exo effect, at 250–400 °C, corresponding to surface oxidation and the transformation proper is not distinct above the Curie point in the 600–1000 °C region. Under nitrogen, the endo effect at 585 °C, corresponding to the Curie point, is more clearly evident.

[17]) Experiments with synthetic samples only; in natural substances FeO is formed at decomposition only and it is immediately oxidized.

[18]) The position and the shape of the exo effect (250—500 °C) depend on pH during the precipitation and on the method of preparation. At higher pH values a broader effect is observed at higher temperatures. The higher temperature of precipitation also shifts

(continued)

	Side effect				
Temperature of the peak T_m °C	Temperaturə of the break-point T_0 °C	Effect	Reaction	Reference	Remarks
590–650	550	exo small	$\gamma \to \alpha\text{-Fe}_2\text{O}_3$	[49], [174]	[16]
215 263		endo small endo small	unidentified	[79]	[17]
120		endo large	dehydration	[10], [114], [49], [88], [98], [104], [105], [160]	[18]
250–300 525	480	endo small endo small shallow	dehydration formation of böhmite γ-AlO . OH decomp. of böhmite	[28], [33], [39] [39]	[19]
850		exo inflection	$\alpha \to \Theta\text{-Al}_2\text{O}_3$		
250–300 525 950 1040	480	endo small endo small exo small exo small	formation of böhmite decomp. of böhmite $\chi - \text{Al}_2\text{O}_3 \to$ $\to \varkappa\text{-Al}_2\text{O}_3$ $\to \alpha\text{-Al}_2\text{O}_3$	[28], [31], [33] [41], [17] [139]	[20]

the effect to higher temperatures. The presence of some anions (SO_4^{2-}, HPO_4^{2-}) affects crystallization and thus also the position of the peak; natural substances show an exo effect at 350–400 °C.

[19]) The curves of various samples are not reproducible in the 250–335 °C temperature range. They create up to three effects; small endo effects in the 250–300° region and at 525 °C correspond to the formation of böhmite and need not be present; the total decomposition mechanism is: $\alpha\text{-Al(OH)}_3 \to \gamma\text{-Al}_2\text{O}_3 \pm \gamma\text{-AlO . OH} \to \gamma\text{-Al}_2\text{O}_3 \to \Theta\text{-Al}_2\text{O}_3 \to \alpha\text{-Al}_2\text{O}_3$

[20]) The curves are well reproducible; the total mechanism of decomposition is:

$$\gamma\text{-Al(OH)}_3 \to \begin{matrix} \chi\text{-Al}_2\text{O}_3 \longrightarrow \varkappa\text{-Al}_2\text{O}_3 \searrow \\ \gamma\text{-AlO . OH} \to \gamma\text{-Al}_2\text{O}_3 \to \delta\text{-Al}_2\text{O}_3 \to \Theta\text{-Al}_2\text{O}_3 \nearrow \end{matrix} \alpha\text{-Al}_2\text{O}_3 .$$

Table 4.5

Substance	Main effect			
	Temperature of the peak T_m °C	Temperature of the break-point T_0 °C	Effect	Reaction
HAlO$_2$ or α-AlO . OH Diaspore (according to the USA system β-monohydrate)	540—595	490	endo large	dehydration
γ-AlO . OH Böhmite (according to the USA system γ-monohydrate)	510–580	450	endo large	dehydration
γ-Al$_2$O$_3$ χ-Al$_2$O$_3$ α-Al$_2$O$_3$	850 950		exo small exo small	→ Θ-Al$_2$O$_3$ → χ-Al$_2$O$_3$
Bauxite (a mixture of böhmite, hydrargillite, and diaspore)	330—580		endo large	dehydration
MnO . OH Manganite	370—380	360	endo sharp	dehydration → Mn$_2$O$_3$
MnO Manganosite	640—650	520	endo small	loss of active oxygen
(Mn, Fe)$_2$O$_3$ Bixbyite	1009		endo small	→ β-Mn$_3$O$_4$
Mn$_3$O$_4$ Hausmannite	950		endo small irreversible	α → β-Mn$_3$O$_4$

[21]) With finely granulated samples the peak temperature is lower; the presence of Mn^{3+} in manganese-diaspore does not affect the curve.

[22]) The particle size affects T_m. Very finely granulated synthetic samples have a peak at 450 °C. A small exo effect at 830—930 °C depends on the crystallinity of the sample; usually contaminated with candite.

[23]) Among anhydrous aluminas only γ and χ manifest themselves by weakly exothermic transformations; in some instances a weak exo effect of the transformation to α-Al$_2$O$_3$ can also be observed at 1000—1100 °C.

[24]) This is not an unambigously defined mineral. It contains the mentioned compo-

(continued)

Side effect				Reference	Remarks
Temperature of the peak T_m °C	Temperature of the break-point T_0 °C	Effect	Reaction		
				[53], [52], [77], [81], [145], [100], [103]	21)
850—930		exo inflection	δ-$Al_2O_3 \rightarrow$ Θ-Al_2O_3	[52], [54], [77], [81], [82], [100], [103], [126]	22)
950				[17]	23)
				[102]	24)
960—980	850	endo small double	Mn_2O_3 $\rightarrow Mn_3O_4$	[10], [15] [89], [127],	25)
550—600			unidentified	[89], [127], [132]	
550—600		endo small			26)
1160—1250		endo small reversible	$\beta \rightarrow \gamma$-Mn_3O_4	[127], [132]	27)
1160—1250		endo small reversible	$\beta \rightarrow \gamma$-Mn_3O_4	[132]	

nents either mixed or separately; iron oxides may accompany it. The possible accompanying effects depend on the composition.

25) The composition of the product above 370 °C depends on the rate and the conditions of heating and Mn_2O_3 may be accompanied by pyrolusite, MnO_2, or Mn_3O_4. This also affects the course of the curve above 370 °C.

26) Applies the same as for manganite. The data given are valid for heating in a vacuum. When heating in air is carried out, oxidation takes place.

27) The transformation in the case of Mn_3O_4 is evident in the curves of all oxides of manganese.

Table 4.5

Substance	Main effect			
	Temperature of the peak T_m °C	Temperature of the break-point T_0 °C	Effect	Reaction
MnO_2 Pyrolusite	620—650	580	endo large	→ Mn_2O_3 (bixbyite)
MnO_2 Ramsdellite	500 650—700	450 610	exo small endo large	→ pyrolusite → Mn_2O_3
KMn_8O_{16} Cryptomelane	900—1000	880	endo medium	→ Mn_3O_4
$BaMg_8O_{16}$ Psilomelane	900—1000		exo small	
$Mg(OH)_2$ Brucite	450	390 340	endo large endo large	→ MgO → MgO
$Ca(OH)_2$ Portlandite	585	480	endo large	→ CaO
SiO_2 Quartz	573	550	endo small sharp, reversible	$\alpha \rightleftharpoons \beta$-transformation

[28]) Other forms of MnO_2 such as α, γ, δ and manganomanganite, are usually contaminated by foreign ions and their curves are very similar.

[29]) The shape of the curve up to 900 °C may be very varied, depending on the source locality, and it displays both endo and exo effects.

[30]) For undiluted sample.

[31]) In a dolomite sample the region of the main peak changes with the composition and the granulation; water vapour and CO_2 shift the main peak to higher temperatures. The fineness of the particles increases the magnitude of the main effect.

[32]) The particle size affects the magnitude of the effect. In CO_2 the reaction is shifted to higher temperatures and the main endo effect is diminished in consequence of a strong carbonate formation; in the case of a mixture of $Ca(OH)_2$ and $Mg(OH)_2$ where $Ca(OH)_2$

(continued)

Side effect				Reference	Remarks
Temperature of the peak T_m °C	Temperature of the break-point T_0 °C	Effect	Reaction		
980—1050	950	endo small	$\rightarrow \beta$-Mn_3O_4	[10], [129], [132]	28)
1200		endo small reversible	$\beta \rightarrow \gamma$-Mn_3O_4		
950—1050		endo small	$\rightarrow \beta$-Mn_3O_4	[89], [132]	
1200		endo small reversible	$\beta - \gamma$-Mn_3O_4		
				[129], [132]	29)
				[89]	
790—840		endo small	residue of dolomite	[116], [144]	30) 31)
				[86], [159]	32)
				[42], [60], [65], [11], [38], [12], [61], [62], [63], [64], [78], [117], [118]	33)

content is above 70% the main peak is followed by a small endo effect at 550 °C.

33) The DTA curve is similar to irreversible reactions with a sharper peak. The deviation from the base line, caused by different diffusivity of the sample and the reference, is usually eliminated by the dilution of the sample and by the use of cooling curves. The constancy of the transformation temperature allows utilization of SiO_2 for calibration. Maximum range, determined from a large number of natural samples, is 572—574 °C. In the case of synthetic samples a much larger scatter was observed (up to 160°), explicable by the anomalous composition of the samples and by the possibility of the formation of solid solutions. The differences of granulation in the —35 to +200 mesh range do not have a substantial influence. The presence of another form of the silicate and the possibility of the change of crystalline character during grinding may have an effect.

Table 4.5

Substance	Main effect			
	Temperature of the peak T_m °C	Temperature of the break-point T_0 °C	Effect	Reaction
SiO_2 Crystobalite	220—260	220	endo small sharp, reversible	$\alpha \rightleftharpoons \beta$-transformation
SiO_2 Tridymite	110—160 110—160		endo small sharp, reversible	$\alpha \rightleftharpoons \beta$-transformation
SULPHATES				
K_2SO_4 Arcanite	583—600	560	endo large sharp	transformation
Na_2SO_4 Thenardite	243	185	endo large	polymorphous transformation abruptly finished at 227 °C
$CaSO_4 \cdot 2 H_2O$ Gypsum	100—170 150—210	80	endo large endo small	dehydration
$Na_2Mg(SO_4)_2 \cdot 4 H_2O$ Blödite	100		endo medium	dehydration

[34]) This is the product of a high-temperature transformation of quartz. As the $\alpha \rightleftharpoons \beta$ transformation takes place at relatively low temperatures, it requires a special technique (metallic block). The DTA curve has a variable character, in the sense that an increase of time or temperature of the treatment leads to the shift of the peak to higher temperatures with the possibility of the formation of a double peak (229 and 254 °C) and the stabilization of the final peak at 248 °C. This phenomenon is probably caused by the formation of a solid solution with the oxides present (CaO, MgO, Al_2O_3 ..), or by the formation of a liquid phase of the alkali metal oxides present. The shift of the peak temperature to higher values as a result of an increase in the amount of impurities shows that they are more soluble in the β form.

(continued)

Side effect				Reference	Remarks
Temperature of the peak T_m °C	Temperature of the break-point T_0 °C	Effect	Reaction		
				[42], [60], [65]	34)
				[61], [65], [114], [134]	35)
1080		endo small sharp	melting	[68]	
885		endo small sharp	melting	[10], [28] [68], [97]	36)
370—380		endo small shallow	transition of soluble an-hydrite to the insoluble form	[1], [2], [54], [68], [77], [85],	37)
1180—1230		endo small sharp	polymorphous change	[93], [121],	
1350		endo small	melting of the eutectic compound $CaSO_4 \cdot CaO$	[132], [161]	
548	820	endo large narrow	formation of solid solution	[28], [132]	
650	640	endo small	unidentified		

35) Similarly to crystobalite, tridymite also forms a simple or double peak, probably as a result of the existence of solid solutions. The fluctuation of the inverse temperature is explained on the basis of the rotation of oxygen atoms. Therefore the magnitude and the width of the effects also deviate and the possibility of a quantitative evaluation is more difficult than in the case of quartz.

36) An anhydrous form of mirabilite; in its presence an additional endo effect of dehydration takes place at 180 °C.

37) Dehydration takes place in two steps, via $CaSO_4 \cdot 1/2\,H_2O$. The exo effects at temperatures above 500 °C correspond to transitions to insoluble forms.

Table 4.5

Substance	Main effect			
	Temperature of the peak T_m °C	Temperature of the break-point T_0 °C	Effect	Reaction
$K_2MgCa_2(SO_4)_4 . 2H_2O$ Polyhalite	100	90	endo medium	
	135		endo medium	dehydration
	157		endo medium	
$KAl_3(SO_4)_2(OH)_6$ Alunite	500—590	500—530	endo medium	decomposition to $K_2SO_4 . Al_2(SO_4)_3$
	880	800	endo medium	decomposition of the component $Al_2(SO_4)_3$
$KFe_3(SO_4)_2(OH)_6$ Jarosite	400	300	endo medium	decomposition to $K_2SO_4 .$ $Fe_2(SO_4)_3$
	705	650	endo medium	decomposition of the component $Fe_2(SO_4)_3$
$SrSO_4$ Celestite	1180		endo medium	transformation
$BaSO_4$ Barite	1180		endo medium	transformation
$LiSO_4 . H_2O$	210	75	endo large	dehydration
$PbSO_4$	885	885	endo small sharp	transformation
$ZnSO_4$	200	50	endo large	dehydration
	350	290	endo large	dehydration
$KAl(SO_4)_2 . 12H_2O$	100—300		endo large complex	dehydration
$MgSO_4 . 7H_2O$	100—400		a series of large endo effects	dehydration

[38]) The small exo effect before the dehydration itself is connected with a rapid change in c_p during dehydration.

(*continued*)

Side effect				Reference	Remarks
Temperature of the peak T_m °C	Temperature of the break-point T_0 °C	Effect	Reaction		
300		endo small broad	transition to solid solution	[28], [132]	38)
530	480	endo medium broad	transformation K_2SO_4		
780	750	exo small sharp	crystallization Al_2O_3	[5], [28], [44], [47],	
850—870	770	endo large	decom position of anhydrous alum	[68], [77] [85], [90]	
1040		endo small	melting		
500		endo small	crystallization Fe_2O_3	[21], [85], [90]	
1600		endo small	melting with simultaneous decomposition	[68]	
				[68]	
600		endo large	reversible change	[68]	39)
1000—1200		gradual transition to the endo side	decomposition	[68]	
1087		endo small	melting		
800—1080		4 complex endo effects	transformation decomposition	[68]	
840		endo medium		[68]	
1070		break to endo side	melting		
1000—1200		endo small broad	decomposition	[68]	

39) For a synthetic sample.

Table 4.5

SULPHIDES

Sulphides usually accompany clay minerals and on DTA curves they are characterized

Substance	Main effect			
	Temperature of the peak T_m °C	Temperature of the break-point T_0 °C	Effect	Reaction
PbS Galena	337	300	exo medium shallow	oxidation interrupted at 412 °C and then continuing further
	517	480	exo medium sharp	
ZnS Sphalerite	480—500	430	exo medium shallow	oxidation of the surface layer
Cu$_2$S Chalcocite	100	90	endo small	$\alpha \rightleftharpoons \beta$-transformation
	520	300	exo large	
FeS$_2$ Pyrite	406—443	310—354	exo large sharp	oxidation
FeS$_2$ Marcasite	433 482 518	370	exo large sharp	oxidation
CuFeS$_2$ Chalcopyrite	450	300	exo large	oxidation
FeAsS Arsenopyrite	530	490	exo medium	oxidation

[40]) The curves are very complex. Oxidation starts at 300 °C, but its course is stepwise. Above 700 °C complications arise owing to the corrosion of the thermocouples.

[41]) The small endo effect with the break-point at 90 °C is slowed down inexplicably at 95 °C. The main exo effect with a sharp peak at 520 °C is characterized by a wave on the ascending and descending side of the effect.

[42]) Decomposition. Its course and the oxidation products are affected by the atmosphere and the weight of the sample. A simple exo effect is obtained with diluted samples. With undiluted samples the curve displays a very complex course in the 354—700 °C interval with a series of exo effects, as a result of oxidation in the cross-section of the sample; in vacuum, pyrite is decomposed to pyrrhotine and sulphur. Above 700 °C the curve can have a very complex course.

(continued)

mainly by exo effects corresponding to their oxidation. These substances are corrosive toward commonly used thermocouples and metallic crucibles.

Side effect				Reference	Remarks
Temperature of the peak T_m °C	Temperature of the break-point T_0 °C	Effect	Reaction		
830	780	endo large	corrosion of the thermocouple	[102],	[40]
900		break-point of an exo reaction	volatilization	[77], [102]	
823	730	endo large	formation of tenorite (CuO)	[102]	[41]
	1000	an endo reaction	eutectic cuprite→tenorite		
1070		endo small	melting	[54], [96], [16], [102], [135]	[42]
				[54], [96]	[43]
770	600	endo large	corrosion of the thermocouple	[102]	[44]
				[16]	[45]

[43]) The complex course of the exo effect is caused by the insufficient access of oxygen to the inner parts of the sample. In inert atmosphere the endo effect appears at 415 °C, corresponding to the transformation of marcasite to pyrite. In air the endo effect between both peaks of oxidation at 433 and 482 °C may correspond to this reaction. Oxidation is faster than in pyrite.

[44]) The basic exo effect with a sharp peak at 450 °C is broad and the ascending and the descending side reflect the irregularities of the oxidation of large particles.

[45]) For samples diluted with Al_2O_3 in the ratio 1 : 15. Under different conditions a greater number of effects is found.

Table 4.5

CARBONATES

DTA curves of carbonates are characterized predominantly by the enodthermic effect of the decomposition which may be combined with the effect of oxidation of the oxides formed. These peaks may be affected by the experimental method used, i.e. by the rate of decomposition, partial pressure of CO_2, properties of the vessels and block used, particle size, etc. Therefore the curves of carbonates may differ very much and a uniform method

Substance	Main effect			
	Temperature of the peak T_m °C	Temperature of the breakpoint T_0 °C	Effect	Reaction
$CaCO_3$	860—1010		endo large	$CaCO_3 \rightarrow$ $CaO + CO_2$
Calcite	830—940	830—920	endo large	$CaCO_3 \rightarrow$ $CaO + CO_2$
Aragonite	830		endo large	$CaCO_3 \rightarrow$ $CaO + CO_2$
$MgCO_3$ Magnesite	660—690	580—630	endo large	$MgCO_3 \rightarrow$ $MgO + CO_2$
$(FeMg) CO_3$ Breunnerite	700—755		endo large	$MgCO_3 \rightarrow$ $MgO + CO_2$
$(FeMg) CO_3$			endo large	$FeCO_3 \rightarrow$ $FeO + CO_2$
Pistomesite	600		double	$MgCO_3 \rightarrow$ $MgO + CO_2$
$MgCO_3 . 3 H_2O$ Nesquehonite	150—250	endo large double		$-2 H_2O$
	400—450		endo medium	$-1 H_2O$
	450—550		endo small double	$MgCO_3 \rightarrow$ $MgO + CO_2$

[46a]) For undiluted samples; [46b]) for samples diluted to 20%.

[47a]) For some samples of calcite and aragonite with two sorts of calcite.

[47b]) Suitable for the identification of aragonite.

[48]) The presence of NaCl (1%) decreases the temperature of the peak by approximately 50 °C.

(continued)

of analysis is recommended: (*a*) adjustment of the sample to uniform particle size, 50—300 μm, (*b*) predrying of the sample, (*c*) analysis in a defined neutral atmosphere or in vacuum (if oxidation is imminent), (*d*) elimination of soluble salts and the dilution of the sample with inert material in special cases, (*e*) experimental technique should be adapted to the liberation of CO_2

Side effect				References	Remarks
Temperature of the peak T_m °C	Temperature of the break-point T_0 °C	Effect	Reaction		
1010 389—488	900	endo small	transformation aragonite→calcite	[10], [66], [76], [36], [37]	46a) 46b) 47a) 47b)
690—700	500	exo small	formation of intermediary $MgO . MgCO_3$ or oxidation in the case of breunnerite	[6], [9], [95], [140], [152] [152]	48)
				[6]	49)
650		exo small	$FeO \rightarrow Fe_2O_3$	[6], [7]	50)
550—600		exo small	crystallization of amorphous MgO	[6], [7]	

49) According to the content of FeO a small exo effect may be observed after the main reaction.

50) The decomposition product is magnesioferrite $(MgFe)Fe_2O_4$.

Table 4.5

Substance	Main effect			
	Temperature of the peak T_m °C	Temperature of the breakpoint T_0 °C	Effect	Reaction
$MgCO_3(OH_2 \cdot 3\,H_2O$ Artinite	280 400—550	240 380	endo medium endo effect combined	dehydration a) dehydration b) —CO_2
$Mg_5(CO_3)_4(OH)_2$ Hydromagnesite	340	200	endo large	dehydration —2 H_2O
	450	400	endo medium	dehydration —2 H_2O
$CaMg(CO_3)_2$ Dolomite	790—940	745—870	exo large double double	a) $MgCO_3 \rightarrow$ $MgO + CO_2$ b) $CaCO_3 \rightarrow$ $CaO + CO_2$
$Ca(MgFe)(CO_3)_2$ Ankerite	700—720 850—860 910—930	700 880	exo large endo medium endo medium	disintegration of the rhombic formation of calcite complexes decomposing to calcite
$CaMg_3(CO_3)_4$ Huntite	690—700 900	570 850	endo large endo medium	$MgCO_3 \rightarrow MgO$ + CO_2 $CaCO_3 \rightarrow CaO$ + CO_2
$FeCO_3$ Siderite	585	500	endo large	$FeCO_3 \rightarrow FeO$ + CO_2

[51]) The presence of small amounts of salts may shift the temperature of the first peak to both sides. In a CO_2 atmosphere MgO reforms $MgCO_3$. The pressure of CO_2 affects the clarity of both peaks. It is best at 760 mm Hg pressure.

[52]) Three endo effects are typical, sometimes also a weak exo reaction.

[53]) Behaves as a mixture of magnesite and calcite.

(continued)

Side effect				Reference	Remarks
Temperature of the peak T_m °C	Temperature of the break-point T_0 °C	Effect	Reaction		
550	510	exo small	crystallization of amorphous MgO	[6]	
500	490	exo medium sharp	crystallization of amorphous MgO	[6]	
570	530	endo medium	decomp. of carbonate		
				[7], [70], [95], [115], [135]	51)
				[95]	52)
				[37]	53)
670	650	exo large	FeO → Fe$_2$O$_3$. (Fe$_3$O$_4$)	[40], [83], [79], [136],	54)
830	800	exo small	γ-Fe$_2$O$_3$ → haematite	[140]	

54) According to the conditions, all three effects may be appreciably variable. The main decomposition reaction is accelerated by the presence of water vapour and it may be superimposed in various ways by the first exo reaction. In CO_2, the exo effect does not occur, in an oxidation medium only an exo effect is observed. The exo effect at 800 °C corresponds to the transformation of the previously formed oxidation product and it is absent if an N_2 medium is used.

Table 4.5

Substance	Main effect			
	Temperature of the peak T_m °C	Temperature of the break-point T_0 °C	Effect	Reaction
$SrCO_3$ Strontianite	1200	1000	endo large	$SrCO_3 \rightarrow SrO +$ CO_2
$BaCO_3$ Witherite	1350	1200	endo large	$BaCO_3 \rightarrow BaO +$ CO_2
$ZnCO_3$ Smithsonite	525	440	endo large	$ZnCO_3 \rightarrow ZnO +$ CO_2
$Zn_5(CO_3)_2(OH)_6$ Hydrozincite	310		endo large	loss of H_2O and CO_2
$(CaZn)CO_3$ Nicholsonite	850—1000		endo large	$CaCO \rightarrow CaO +$ CO_2
$(ZnCu)_5(CO_3)_2(OH)_6$ Aurichalcite	275		endo medium	dissociation of hydroxide
$PbCO_3$ Cerussite	330—400	270—340	endo large	$PbCO_3 \rightarrow PbO \,.$ $PbCO_3$
$Pb_3(CO_3)_2(OH)_2$ Hydrocerussite	330—400	270—340	endo large	$PbCO_3 \rightarrow PbO \,.$ $PbCO_3$
$MnCO_3$ Rhodochrosite	600—635	540	endo large	$MnCO_3 \rightarrow MnO$ CO_2

[55]) The phase transition (rhombic-trigonal) is reversible, during cooling the effect at 850 °C is usually masked by the endo effect of calcite.

[56]) In the presence of $BaCa(CO_3)_2$, bromlite and baritocalcite, additional weak endo effects were observed at 585 and 650 °C.

[57]) With very finely granulated samples a weak endo effect at 270 °C was observed.

[58]) The end product is a mixture of ZnO and CuO.

(continued)

Temperature of the peak T_m °C	Temperature of the break-point T_0 °C	Side effect		Reference	Remarks
		Effect	Reaction		
930		endo medium	phase transition	[6], [29], [66]	[55]
830		endo small	$\alpha \rightleftharpoons \beta$-transformation	[6], [29], [66]	[56]
975			$\beta \rightleftharpoons \gamma$-transformation		
				[6], [7], [29]	[57]
				[6]	
500		endo small	$ZnCO_3 \rightarrow$ $ZnO + CO_2$	[6]	
300		inflection on the endo effect	decomposition of carbonate	[6]	[58]
390—440	370	endo small	$PbO \cdot PbCO_3 \rightarrow$ PbO	[6], [29], [66]	[59]
660—850		endo small	volatilization and melting		
300		endo medium	loss of H_2O	[6]	
390—440		endo small	PbO		
660—850		endo small	volatilization and melting		
670		exo large	$MnO \rightarrow Mn_3O_4$	[7], [10], [15], [30], [87], [94]	[60]

[59]) The spread of the results is appreciable; in the 290—400 °C interval a three-step decomposition via the oxycarbonate step was sometimes observed.

[60]) The similarity with the curves of chalybite makes identification impossible. The main effect is often split by the formation of oxycarbonate and by the two-step oxidation of MnO; the spread of T_m values of the main peak is caused by the exchange for Mn^{2+} ion or by impurities; the atmosphere affects the formation of an exo effect.

Table 4.5

Substance	Main effect			
	Temperature of the peak T_m °C	Temperature of the break-point	Effect	Reaction
$Cu_3(CO_3)_2(OH)_2$ Azurite	430	380	endo large	decomposition
$Cu_2CO_3(OH)_2$ Malachite	430	380	endo large	decomposition
$(BiO)_2CO_3$ Bismuthite	500		endo large	decomposition
$NaHCO_3$ Nahcolite	205	150	endo large	decomposition
$KHCO_3$	235		endo large	decomposition
$(NH_4)_2CO_3$	150		endo large	decomposition
Li_2CO_3				

SILICATES

(a) Kaolin minerals

The DTA curves of all kaolin minerals are characterized by distinct endo effects at 100—150 °C, 500—600 °C, and 930 °C, and by two exo effects at 950—980 °C and 1200—1300 °C. The second endo effect, representing the main effect of dehydration, is characterized by its symmetry, which serves for diagnostic purposes. The first exo effect is large and sharp and it characterizes the given group. Views concerning the reactions corresponding to both effects differ strongly and they are

$Al_4(Si_4O_{10})(OH)_8$ Nacrite	660	580	endo large symmetrical	dehydration
$Al_4(Si_4O_{10})(OH)_8$ Dickite	660—690	580	endo large asymmetrical	dehydration

[61]) Up to the melting temperature no heat of reaction is observed.

[62]) The published curves display a different course of dehydration and discontinuity thus demonstrating contamination with other minerals; the course of dehydration is

(continued)

Side effect				Reference	Remarks
Temperature of the peak T_m °C	Temperature of the break-point T_0 °C	Effect	Reaction		
				[6]	
				[6], [7]	
600		endo small	transformation	[6]	
750—790		endo small irregular		[6] [6]	
				[6], [7]	
				[6], [7]	
> 700		endo irregular	melting	[7]	[61])

discussed in detail by Mackenzie [102], as is also the effect of granulation, crystallinity, impurities, dehydration kinetics, etc. Not all effects mentioned need occur in various samples. The presence of the first large endo effect indicates allophane or halloysite. The first small endo effect indicates partly the dehydrated halloysite or a non-arranged mineral. The absence of the first endo effect indicates nacrite, dickite, or coarsely granulated kaolinite.

950—980	950	exo medium sharp	structural change	[27], [84], [85], [133]	[62])
950—980	950	exo medium sharp	structural change	[27], [84], [85], [133]	

dependent on the granulation and the crystallinity of the sample. Dickite displays an endo effect at a temperature approximately 20 °C higher, which makes a double effect possible in its presence.

Table 4.5

Substance	Main effect			
	Temperature of the peak T_m °C	Temperature of the break-point T_0 °C	Effect	Reaction
$Al_4(Si_4O_{10})(OH)_8$	100—200		endo very small	loss of adsorbed water
Kaolinite	600	510	endo large symmetrical	dehydration
Kaolinite pM	100—200		endo small	loss of adsorbed
	500—580	430	endo large asymmetrical	dehydration
$Al_4(Si_4O_{10})(OH)_8 \cdot$ 4 H_2O	150	100	endo medium	loss of adsorbed water
Halloysite	570		endo large	dehydration
$(Al_4H_3)_4(OH)_8Si_4O_{10}$	100—120		endo small	loss of adsorbed water
Anauxite	600	510	endo large asymmetrical	dehydration
$m.Al_2O_3.nSiO_2.pH_2O$	140—150	100	endo large	dehydration
$m.Fe_2O_3.nSiO_2.pH_2O$ Hisingerite	120	100	endo medium	loss of adsorbed water

[63]) The presence of fine fractions causes the formation of endo effects at 100—200 °C. In the case of pure samples a small endo effect at 930 °C appears.

[64]) The effect at 100—200 °C is always present; the effect at 500—600 °C is not symmetrical as in the previous case; the exo effect at 980 °C is narrower.

[65]) In the case of a less hydrated sample the first endo effect is usually smaller.

(continued)

Side effect				Reference	Remarks
Temperature of the peak T_m °C	Temperature of the break-point T_0 °C	Effect	Reaction		
950—980	930	exo large sharp	structural change	[50], [55], [102]	[63]
950—980	920	exo large narrow	structural change	[55], [122]	[64]
950—980	930	exo medium	structural change	[55], [122]	[65]
980		exo medium sharp	structural change	[84], [122]	[66]
950	900	exo medium sharp	structural change	[55], [133], [148], [149]	[67]
470	450	exo small shallow		[150]	[68]

[66] The curve has the same course as that of kaolinite.

[67] The major part of the water has evaporated at 200 °C; the degree of rearrangement determines the magnitude of the exo effect at 900—1000 °C.

[68] The course of the curves is analogous to those of allophane, Al_2O_3 is substituted by Fe_2O_3.

Table 4.5

(*b*) Montmorillonite minerals

DTA is difficult, as the temperature effects are usually small and variable. Often the baseline is irregular owing to their ability to sorb organic substances and to expand single structural units. Work with them requires careful standardization of the method and usually also pretreatment of the sample. The DTA curve may be divided into two parts, i.e. the region of the absorbed water (100—300 °C) and the region of the loss of the bound water (500—1000 °C) which is usually accompanied by a change of the structure and recrystallization. The magnitude and the character of the endo effect and its position on the temperature axis in the first region are usually not connected with the mineralogical character of the sample but with its pretreatment, with respect to the possibility of adsorption of exchangeable cations Mg^{2+}, Ca^{2+}, Ba^{2+}, Li^+, Na^+, NH_4^+, Rb^+, Be^{+2}, Cs^+ and the moisture. When cations Mg^{2+}, Ca^{2+}, Sr^{2+}, H^+ are hydrated, expansion of the bimolecular layer of water takes place if the degree of hydration is low. In

Substance	Main effect			
	Temperature of the peak T_m °C	Temperature of the breakpoint T_0 °C	Effect	Reaction
$m(Mg_3)(Si_4O_{10})$ $[(OH)_2 p(Al, Fe)]_2$ $(Si_4O_{10})(OH)_2$ Montmorillonite	180⎱ 250⎰	100	endo medium	loss of adsorbed water
	150 240	100	endo medium doubled	loss of adsorbed water
	160	100	endo medium	loss of adsorbed water
	120	100	endo medium	loss of adsorbed water
$Al_{2.17}[(OH)_2Al_{0.83}Si_{3.17} \cdot O_{10}]^{0.32} Na_{0.32}(H_2O)_4$ Beidellite	120	100	endo medium doubled	loss of adsorbed water
$Fe_2^{3+}[(OH)_2Al_{0.35}SiO_{0.67}O_{10}]^{0.35-} Na_{0.33}(H_2O)_4$ Nontronite	150 200	100	endo medium doubled	loss of adsorbed water
$Mg_3[(OH)_2Al_{0.33}Si_{3.67}O_{10}]^{0.33-} Na_{0.33}(H_2O)_4$ Saponite	150 200	100	endo medium doubled	loss of adsorbed water

[69]) Sample with adsorbed Mg^{2+}.
[70]) Sample with adsorbed Ca^{2+}.
[71]) Sample with adsorbed Na^+.

(continued)

the case of cations Ba^{2+}, Li^+, Na^+ and K^+ a monomolecular layer is formed. When cations Mg^{2+}, Ca^{2+}, H^+, are adsorbed a double endo effect is observed, when Na^+, NH_4^+, Rb^+, Be^{2+}, are adsorbed only a simple effect is observed. In the second region (500—1000 °C) the position of the endothermic peak depends primarily on the amount of Fe^{3+} present in the sample, which isomorphically substitutes Al^{3+}. In the case of small samples the endo effect lies in the 450—750 °C region with the peak at 700 °C. With increasing content of iron the temperature of the endo effect peak decreases (550—600 °C) for nontronite. In the 800—900 °C region some samples show endothermic reaction corresponding to a complete destruction of the crystal structures or to the escape of water bound to Mg^{2+}. The subsequent exothermic reaction leads to the formation of variable products (mullite, crystobalite, cordierite, periclase, spinel). The DTA of these minerals is very difficult and a series of methods for the pretreatment of the sample has therefore been developed.

Temperature of the peak T_m °C	Temperature of the break-point T_0 °C	Side effect		Reference	Re-marks
		Effect	Reaction		
650⎱⎰ 820	580	endo large doubled	loss of structural water	[3], [55], [17], [56], [1], [35], [72], [106], [112], [124], [113], [153]	[69]
890		exo medium			
650	580	endo large	loss of structural		[70]
810		doubled	water		
610	550	endo large	loss of structural		[71]
790		doubled	water		
610	550	endo large	loss of		[72]
820		doubled	structural water		
550		endo large	loss of structural water	[102]	
500		endo large	loss of structural water	[102]	
800—900		endo large	loss of structural water	[102]	[73]

[72] Sample with adsorbed K^+.
[73] Hektorite behaves in the same manner.

Table 4.5

(c) Mica minerals

DTA curves of mica minerals may be divided into three characteristic temperature intervals. The interval 50—250 °C is characterized by an endo effect with peak at 125 °C, corresponding to the loss of hygroscopic moisture. This effect is always absent on the curves of pyrophyllite, mica, sericite, and coarsely granulated muscovite. The medium endo effect in this region is evident on curves of glauconite, gadolinite, and finely granulated muscovite. Other mica minerals display a large endo effect in this region. The temperature interval 250—700 °C is characterized by the main endo effect in the 450—650 °C interval with peak at 550 °C which corresponds to the loss of structurally

Substance	Main effect			
	Temperature of the peak T_m °C	Temperature of the breakpoint T_0 °C	Effect	Reaction
$Al_2[(OH_2)_2Si_4O_{10}]$ Pyrophyllite	775	650	endo large shallow	loss of structural water
$Mg_3(OH)Si_4O_{10}$ Mica	1000	870	endo medium	loss of structural water
$KAl_2(OH)_2AlSi_3O_{10}$ Muscovite	750–890		endo medium variable	loss of structural water
$KAl_2(OH)_2AlSi_3O_{10}$ Sericite	700	550	endo medium variable	loss of structural water
$(K, H_2O)Al_2(H_2O, OH)_2 \cdot AlSi_3O_{10}$ Illite	130 550–575	100	endo medium endo medium broad	loss of adsorbed water loss of structural water
$(K, Ca, Na)(Al, Fe^{2+}, Mg)_2 \cdot (OH)_2Al_{0.35}Si_{3.65}O_{10}$ Glauconite	140 550–650	100	endo medium endo medium	loss of adsorbed water loss of structural water
$(K, Ca, Na^+)_1(Al, Fe^{3+}, Fe^{2+}, Mg)_2 \cdot [(OH)_2Al_{0.11}Si_{2.4}O_{10}]$ Seladonite	125 600	500	endo small endo large asymmetrical	loss of adsorbed water loss of structural water

[74]) Temperature T_m is affected by crystallinity and the peak shape by the particle size.

[75]) The particle size affects the shape of the peak appreciably. A finely granulated sample gives an endo effect with the peak at 850 °C.

[76]) The peak of the endo effect is shifted to a lower temperature when the particle size diminishes.

(continued)

bound water. Structural changes take place above 700 °C. The 800—1000 °C interval contains structural changes of anhydrous sample which manifest themselves by a third endo effect at about 900 °C. In the 900—1000 °C interval a small exo effect appears connected with the formation of spinels. The greater the Mg^{2+} content the greater is this effect. In the region about 950 °C a glass amorphous phase may be formed from the alkalis and SiO_2 present, which affects the width and the shape of the third endo effect. The end product of the decomposition may be various (spinel, Al_2O_3, α-Al_2O_3, γ-Al_2O_3, glass, olivine, leucite etc.).

Side effect				Reference	Re-marks
Temperature of the peak T_m °C	Temperature of the break-point T_0 °C	Effect	Reaction		
1200		exo shallow	crystallization to mullite	[84]	[74]
940		endo small	recrystalization to leucite	[57], [107]	[75]
				[57], [107]	[76]
900		endo-exo inversion	exchange Al^{3+} for Mg^{2+} and Fe^{3+}	[58], [99], [109], [108], [119]	[77]
950				[102]	[78a]
950		endo-exo inversion	transformation		
925—950		endo-exo inversion	transformation	[102]	

[77] Samples of various origin change appreciably, mainly with respect to the position of the main endo effect which may be in the 550—720 °C temperature interval.

[78a] The first endo effect at 140 °C need not be present.

Table 4.5

Substance	Main effect			
	Temperature of the peak T_m °C	Temperature of the breakpoint T_0 °C	Effect	Reaction
$Mg_3[(OH)_2Si_4O_{10}]$ Talc	950⁻100	870	endo medium	loss of structural water

[78b]) The effect is usually free from interferences of contaminating substances.

On the basis of thermogravimetric curves mica minerals may be divided into the following groups:

(a) Samples displaying small weight losses up to 350 °C, dehydration can be observed above 500 °C. Well crystalline samples with low substitution in the octahedral lattice (coarsely granulated muscovite, sericite, pyrophyllite, mica).

(b) Samples displaying medium losses of weight up to 350 °C and a series of inflexions

---

(d) Vermiculite minerals

The DTA curves of the vermiculite minerals may be divided into three temperature regions; low-temperature interval (0—350 °C) in which dehydration takes place, usually in several steps according to the lamination of minerals and the amount of interlaminar water. In the case of vermiculite saturated with Mg^{2+}, this dehydration takes place in three steps which are superimposed in various ways. The numbers of H_2O molecules per cation, liberated during this dehydration, are approximately 4, 5 and 2.5. The remaining water (approx 3/4 mole per cation) departs only at higher temperatures. In the case of samples saturated by other cations (not Mg), for example Ca^{2+}, Sr^{2+}, Ba^{2+}, or Na^+, the differences in the hydrating system are known, but cannot be identified easily by DT

$(Mg, Fe^{2+}Fe^{3+})_3[(Si, Al)_4O_{10}]$. $(OH)_2$. $6 H_2O$	130 175 265	80	endo large triple	loss of free interlaminar water
Mg Vermiculite	540		endo small shallow	loss of the residue of free water
$(Mg, Al)_3[(OH)_2AlSi_3O_{10}]^{0.33}$ − $Mg_{0.33}(H_2O)_4$ Batavite	130 175 265	80	endo large triple	loss of free interlaminar water

[79]) Endo effects of dehydration are superimposed in various manners, depending on the amount of moisture present before the analysis.

[80]) On saturation with cation the peaks of the endo effects of free H_2O are shifted

(continued)

Side effect				Reference	Re- marks
Temperature of the peak T_m °C	Temperature of the break- point T_0 °C	Effect	Reaction		
				[102]	78b)

of subsequent dehydration, between 350 and 600 °C (glauconite, finely granulated muscovite)

(c) Samples displaying approximately a 5% weight loss up to 350 °C and a distinct inflexion in the 350—500 °C region (illite).

(d) Samples not giving inflexions, but showing a continuous loss of weight. (Corresponds to complete breakdown of the muscovite lattice. Muscovite ground in dry state for 24 hours).

A in view of the mutual superimposition of the effects. The samples saturated with cations NH^+, K^+, Rb^+, and Cs^+ display in this low-temperature region only a very small endo effect. In the medium temperature interval (350—700 °C) the sample saturated with Mg^{2+} undergoes the loss of the remaining 3/4 mole of water. The small shallow endo effect due to this dehydration is usually distorted by a weak exo effect of the oxidation of FeO present in some samples. In the 700—1000 °C region a two-step loss of structurally bound water takes place. Between both these endo effects an exo effect is observed due to recrystallization to enstatite. The second of the mentioned endo effects is affected appreciably by the saturating cations.

800—810	775	endo medium sharp	partial loss of structural water	[3], [4], [17], [75], [91]	79)
840	810	exo medium sharp	recrystallization to enstatite		80)
850—860		endo small	loss of the residual structural water		
810—940	800	endo small shallow	loss of structural water	[102]	81)

to lower values in the following sequence: $Mg^{2+} > Ca^{2+} > Sr^{2+} > Li^+ > Ba^{2+} > Na^+$. The strength of the aqueous layer is dependent on the type of cations.

81) The exo effect of recrystallization has not been observed.

Table 4.5

(e) Chlorite minerals

Characteristic for chlorite minerals is a sharp endo effect on the DTA curves, with peak at 630 °C, and a small endo effect at 830 °C, followed immediately by an exo effect of various magnitude at 900 °C. Various samples show a doubling of one or both of the mentioned endo effects, DTA and TG curves are strongly dependent on the physical

Substance	Main effect			
	Temperature of the peak T_m °C	Temperature of the breakpoint T_0 °C	Effect	Reaction
$(Mg, Fe, Al)_3(OH)_2Al_{1.2-1.5} \cdot$	110	90	endo small	loss of free water
$Si_{2.8-2.5}O_{10}Mg_3(OH)_6$	510–630	600	endo large ⎫	
Rhipidolith	740		endo small ⎬	loss of bound water
	770		endo small ⎪	
	820		endo small ⎭	
$(Mg, Al)_3[(OH)_2Al_{1.2-1.5} \cdot$	630–670	510	endo large	loss of bound
$Si_{2.8-2.5}O_{10}]Mg_3(OH)_6$	840–890	780	endo large	water
Sheridanite				
Leuchtenbergite	110	90	endo small	loss of free water
	570–630	560	endo large ⎫	loss of bound water
	710–820	680	endo medium ⎬	
	850–900	800	endo small ⎭	
$(Mg, Al)_3(OH)_2Al_{0.5-0.9} \cdot$	110	90	endo small	loss of free water
$Si_{3.5-3.1}O_{10}Mg_3(OH)_6$				
Pennine	400	320	endo small ⎫	loss of bound water
	570–640	580	endo large ⎬	
	820–860	757	endo large ⎭	
$(Mg, Al)_3(OH)_2AlSi_3O_{10}Mg_3 \cdot$	120	100	endo small	loss of free water
$(OH)_6$				
Clinochlore	570–630	580	endo large ⎫	loss of structural water
	810	800	endo medium ⎬	
	870		endo large ⎭	

[82]) The shape of the curve above 700 °C becomes more complex with increasing particle size.

[83]) With decreasing particle size the second effect decreases and the magnitude of the exo effect increases, while both are shifted to lower temperatures.

[84]) The diminution of the particle size makes the course of the curve more distinct but another endo effect is formed with a peak at 705 °C. The first endo effect and exo effect increase with diminishing particle size.

(continued)

state of the parts in which they occur, mainly on granulation. In a ground sample the curve becomes simpler, but a small endo effect is formed at 110 °C corresponding to the hygroscopic moisture, other effects are somewhat shifted (approx. by 50 °C) to lower temperatures.

Side effect				Reference	Re-marks
Temperature of the peak T_m °C	Temperature of the break point T_0 °C	Effect	Reaction		
750—800		exo small		[92], [74], [46], [59], [18], [19], [123], [120]	82)
880—980		exo small		[102]	83)
900		exo large		[102]	84)
860		exo large sharp		[102]	85)
890		exo large		[102]	86)

85) The second endo effect (820 °C) increases with the fineness of the particles, while the exo effect decreases. At the same time a small endo effect at 400 °C is also observed.

86) A decrease in particle size increases the second endo effect (820 °C) and the exo effect.

Table 4.5

(*f*) Serpentine minerals

In view of their compact structure the amount of hygroscopic moisture is small. The loss of the structural water is evident on the TG curves as a single wave in the region of 600 °C. Endo effects on the DTA curves have for all members of this group a similar character, exo effects differ depending on composition. After a small endo effect at 100 °C (hygroscopic moisture) a large endo effect is formed in the 600—800 °C region, with peak at 650—750 °C. In the case of forms containing Mg^{2+} Fe^{2+}, and Ni^{2+}, a sharp exo effect is formed with peak in the 830 °C region, while the

Substance	Main effect			
	Temperature of the peak T_m °C	Temperature of the breakpoint T_0 °C	Effect	Reaction
$Mg_6(OH)_8Si_4O_{10}$	105	80	endo small	loss of hygroscopic water
Antigorite	710	610	endo large	loss of structural water
Bowenite	100	80	endo small	loss of hygroscopic water
	630–700		endo large broad	loss of structural water
Chrysotile	100	80	endo small	loss of hygroscopic water
	710	600	endo small	loss of structural water
Serpentine	150	100	endo medium	loss of hygroscopic water
	680	550	endo large	loss of structural water
Nepouite	140	100	endo large	loss of hygroscopic water
	640–700	400	endo large double	loss of structural water

[87]) The presence of Fe_2O_3 shifts the main endo effect to lower temperatures and diminishes the exo effect.

(continued)

forms containing Fe^{2+}, Fe^{3+} give an exo effect in the 660 °C region. The main endo effect corresponds to the loss of structural water. The position of this endo effect is dependent on the composition of the sample. The residue after dehydration has the tendency to recrystallize to forsterite or enstatite, which is connected with the formation of the exo effect mentioned. The presence of some impurities, mainly Fe oxides, may affect the recrystallization in both directions.

Side effect				Reference	Re-marks
Temperature of the peak T_m °C	Temperature of the break-point T_0 °C	Effect	Reactions		
850	800	exo large sharp	recrystallization to forsterite	[22], [23], [24]	
				[102]	
850	830	exo small sharp	recrystallization to forsterite	[102]	
840	810	exo large sharp	recrystallization to enstatite	[102]	87)
820	800	exo medium	recrystallization to forsterite	[102]	88)

88) The samples of nepouite display a large scatter with respect to the position of the main endo effect, sometimes amounting to 200 °C.

Table 4.5

(g) Sepiolite and palygorskite minerals

Substance	Main effect			
	Temperature of the peak T_m °C	Temperature of the breakpoint T_0 °C	Effect	Reaction
$4\,MgO\,.\,6\,SiO_2\,.\,2\,H_2O$	100–150	80	endo medium double	dehydration
	350	300	endo small ⎫	loss of bound
	510	450	endo small ⎭	water
$(Mg, Fe^{2+}Fe^{3+})_4(H_2O)_3\,.$ $(OH)_2\,SiO_6O_{15}\,.\,3\,H_2O$ Xylolith	150	80	endo large broad	dehydration
	350–400		endo small ⎫	loss of bound
	510		endo small ⎭	water
$Mg_5Si_8O_{20}(OH)_2(OH_2)_4\,.$ $4\,H_2O$ Palygorskite	130	90	endo large ⎫	
	290	250	endo large ⎬	dehydration
	520	450	endo large ⎭	

(h) Other silicates

$KAlSi_2O_6$	611	580	endo small	transormation rhombic to
Leucite	633		double	cubic
$NaCa_2Si_3O_8OH$ Pectolite	780	744	endo large sharp	dehydration
$Ca_{10}(Mg, Fe)_2Al_4\,.\,Si_9O_{34}(OH)_4$ Idocrase		570–845 T_0-T_{end}	endo shallow irregular	dehydration

[89]) The magnitude of the endo-exo inversion in the 750—880 °C region is dependent on the origin of the sample.

[90]) The endo-exo inversion is less distinct than in the case of sepiolite; it is also shifted to lower temperatures. The doubling of the first endo effect is probably connected with a double binding of water.

(continued)

Side effect				Reference	Re-marks
Temperature of the peak T_m °C	Temperature of the break-point T_0 °C	Effect	Reaction		
830	750	endo large sharp	loss of structural water		
850—880		exo large sharp	recrystallization to enstatite	[84], [10], [11]	89)
720		endo large shallow	loss of structural water	[51]	90)
810		exo small shallow	recrystallization to enstatite		
550	540	endo small sharp	change of structure	[84], [10], [10]	
				[102]	91)
1050		endo small	melting	[102]	
850		exo inflection	recrystallization of amorphous components formed by decomposition (phosphates)	[102]	92)
	1017		melting		

91) The rate of change is small and the doubling of the effect may be caused, therefore, by the presence of particles of various sizes. After the transformation the baseline was shifted.

92) The curves of various samples are poorly reproducible. The composition is also variable, depending on the content of Mg, Ca, Fe and Al.

Table 4.5

Substance	Main effect			
	Temperature of the peak T_m °C	Temperature of the breakpoint T_0 °C	Effect	Reaction
$Ca_2Al_3Si_3O_{12} \cdot OH$ Zoisite	998	938	endo medium sharp	dehydration
$Ca_2(Al, Fe)_3Si_3O_{12} \cdot OH$ Epidote	1018	912	endo medium stepwise	dehydration
$Ca_2Al_2Si_3O_{10}(OH)_2$ Prehnite	787 868	710 820	endo medium	dehydration
$Zn_4Si_2O_7(OH)_2 \cdot H_2O$ Hemimorphite	468 705	380 670	endo small broad endo large sharp	dehydration dehydration
$CuSiO_3 \cdot 2 H_2O$ Chrysocolla	80	50	endo medium broad	dehydration
$Ca_2(Mg, Fe)_5[(OH)Si_4O_{11}]_2$ Amphiboles (Tremolite)	1040	950	endo medium broad	dehydration and formation of pyrocene
Monothermite	555		endo medium sharp	

[93]) On the descending side of the endo effect inflections appear at 938 and 987 °C indicating the presence of zoisite.

[94]) Ascending side of the exo effect shows a step at 904 °C.

[95]) In view of the variable composition, the DTA curves are very poorly reproducible.

(continued)

Side effect				Reaction	Re-marks
Temperature of the peak T_m °C	Temperature of the break-point T_0 °C	Effect	Reference		
				[102]	
				[102]	93)
870		hint of exo effect	recrystallization (Fe_2O_3)	[121]	
1280	1225	endo medium	melting		
937	876	exo small	$Zn_2SiO_4 \rightarrow$ willemite	[39], [77]	94)
290		exo small broad	$CuO + SiO_2$	[77], [151]	95)
690		exo small	impurities		
840	750	exo medium	crystalline contraction	[125], [132], [162]	96)
190		exo small		[8], [13], [52], [141], [142]	97)

96) The effects mentioned concern tremolite. The effect at 840 °C is typical of the majority of amphiboles and in the case of tremolite it is reversible.

97) The name is used mainly in Russian literature; the composition is uncertain.

REFERENCES

1. ALLAWAY W. H. *Proc. Soil Sci. Soc. Am.* **13**, 183 (1949).
2. BARSHAD I. *Am. Mineralogist* **37**, 667 (1952).
3. BARSHAD I. *Am. Mineralogist* **35**, 225 (1950).
4. BARSHAD I. *Am. Mineralogist* **33**, 655 (1948).
5. BAYLISS N. S., KOCH D. F. A. *Australian J. Appl. Sci.* **6**, 298 (1955).
6. BECK C. W. *Am. Mineralogist* **35**, 985 (1950).
7. BECK C. W. *Thesis*, Harvard University 1946.
8. BELYANKIN D. S. *Dokl. Akad. Nauk SSSR* **18**, 673 (1938).
9. BERG L. G. *Dokl. Akad. Nauk SSSR* **49**, 648 (1945).
10. BERG, L. G., NIKOLAEV A. V., RODE E. J. *Termografiya*. Izd. Akad. Nauk SSSR Moscow, 1944.
11. BERKELHAMER L. H. *Tech. Pap. Bur. Mines, Washington* No 664, 38 (1945).
12. BERKELHAMER L. H. *U.S. Bur. Mines Rept. Invest.* No 3763 (1944).
13. BELYANKIN D. S. *Zapiski Vsesoyuz. Mineral. Obshch.* **71**, 1 (1942).
13a. BENARD I., HERTZ I., YEANNIN Y., MOREAU I. *Mémoires Sci. Rev. Métallurg.* LVII. No 5, 389 (1960).
14. BLANC L. *Anal. Chim. Paris* **6**, 182 (1926).
15. BLAŽEK A. *Collection Czech. Chem. Commun.* **25**, 2419 (1960).
16. BLAŽEK A., ČÁSLAVSKÁ V., ČÁSLAVSKÝ J., VÍSAŘ V. *Zpracování Chvaletické rudy.* Research Rept., Mining Institute, Czechslov. Acad. Sci., Prague 1959.
16a. BLAŽEK A. *Proc. Anal. Chem. Conf. Budapest* (1966), 379.
17. BRADLEY W. F., GRIM R. E. *Am. Mineralogist* **36**, 182 (1951).
18. BRINDLEY G. W., ROBINSON K. *X-Ray Identification and Crystal Structures of Clay Minerals.* (Edited by Brindley G. W.) Mineral. Soc., London, 1951, Chapter 6, p. 173.
19. BRINDLEY G. W., ALI S. Z. *Acta Cryst.* **3**, 25 (1950).
19a. BUDNIKOV P. P., SAKHAROV G. P., MATROSOVICH A. N. *Dokl. Akad. Nauk SSSR* **181**, 1430 (1968).
19b. BUDNIKOV P. P., SOLOGUBOVA O. M. *Silikattechnik* **4**, 503 (1953).
20. BROWN J. F., CLARK D., ELLIOT W. W. *J. Chem. Soc.* 84 (1953).
20a. BURNETT D., CLINTON D., MILLER R. P. *I. Mater. Sci* **3**, 1 (1966).
21. CAILLERE S., HÉNIN S. *Bull. Soc. Franc. Minéral. Crist.* **77**, 479 (1954).
22. CAILLERE S., HÉNIN S. *Ann. Agron. (Paris)* **17**, 23 (1947).
23. CAILLERE S., HÉNIN S. *Verre Silic. Industr.* **13**, 114 (1948).
24. CAILLERE S. *Bull. Soc. Franc. Minéral. Crist.* **59**, 163 (1936).
25. CAILLERE S., HÉNIN S. *Compt. Rend.* **221**, 566 (1945).
26. CARR K., GRIMSHAW R. W., ROBERTS A. L. *Trans. Brit. Ceram. Soc.* **51**, 334 (1952).
27. CLARINGBULL G. F. *Mineral. Mag.* **29**, 973 (1952).
28. COCCO G. *Period. Miner.* **21**, 103 (1952).
28a. COPELAND M. S., GOODRICH D. J. *J. Less-Common Metals* **19**, 347 (1969).

29. CUTHBERT F. L., ROWLAND R. A. *Am. Mineralogist* **32**, 111 (1947).
30. TSVETKOV A. I., VAL'YASHIKHINA E. P. *Trudy Inst. Geol. Nauk, Petr. Ser.* **45**, 30 (1955).
30a. ČÁSLAVSKÁ V., BLAŽEK A. *Proc. Anal. Conf. Budapest* 1966, 387.
31. DAY M. K. B., HILL V. J. *J. Phys. Chem.* **57**, 946 (1953).
32. DEB B. C. *J. Soil Sci.* **1**, 212 (1950).
33. DE BOER J. H., FORTUIN J. M. H., STEGGERDA J. J. *Proc. Acad. Sci. Austr.* **B 57**, 170 (1954).
34. DUVAL C. *Inorganic Thermogravimetric Analysis,* 2nd Ed. Elsevier, London 1963, pp. 148–162.
35. EARLEY J. W., MILNE I. H., McVEAGH W. J. *Am. Mineralogist* **38**, 770 (1953).
36. FAUST G. T. *Am. Mineralogist* **35**, 207 (1950).
37. FAUST G. T. *Am. Mineralogist* **38**, 4 (1953).
38. FAUST G. T. *Am. Mineralogist* **33**, 337 (1948).
39. FAUST G. T. *Am. Mineralogist* **36**, 795 (1951).
40. FREDERICKSON J. A. *Am. Mineralogist* **33**, 372 (1948).
41. FRICKE R., HÜTTIG G. F. *Handbuch der allgem. Chemie,* Vol. 9. Hydroxyde und Oxyhydrate, Akademische Verlag, Leipzig, 1937.
42. FENNER C. N. *Am. J. Sci.* **36**, 331 (1913).
43. FRONDEL C. *Mineral. Mag.* **29**, 34 (1950).
44. GAD G. M. *J. Am. Ceram. Soc.* **33**, 208 (1950).
45. GAD G. M., BARRETT L. R. *Mineral. Mag.* **28**, 587 (1949).
46. GALLITELLI P. *Mineral. Petrogr. Mitteil.* **4**, 283 (1954).
47. GEDEON T. *Acta Geol. Hung.* **3**, 27 (1955).
48. GLEMSER O., GWINNER E. *Z. Anorg. Allgem. Chem.* **240**, 161 (1939).
49. GHEIT M. A. *Am. J. Sci.* **250**, 677 (1952).
50. GLASS H. *Am. Mineralogist* **39**, 193 (1954).
51. GREENE-KELLY R. *Clay Minerals Bull.* **2**, 79 (1953).
52. GORBUNOV N. I., STURYGINA E. A. *Pochvovedenie (Pedolog.)* **6**, 357 (1950).
53. GRIM R. E., ROWLAND R. A. *Am. Mineralogist* **27**, 746, 801 (1942).
54. GRIM R. E., ROWLAND R. A. *J. Am. Ceram. Soc.* **27**, 65 (1944).
55. GRIM R. E. *Clay Mineralogy.* McGraw-Hill, New York, 1953.
56. GRIM R. E., BRADLEY W. F. *Am. Mineralogist* **33**, 50 (1948).
57. GRIM R. E., BRADLEY W. F., BROWN G. *X-ray Identification and Crystal Structures of Clay Minerals.* (Edited by Brindley G.W.) Mineral. Soc., London, 1951, Chapter 1 p. 138.
58. GRIM R. E., BRAY R. H., BRADLEY W. F. *Am. Mineralogist* **22**, 813 (1937).
59. GRIM R. E. *Clay Mineralogy.* McGraw-Hill, New York, 1953, p. 198.
60. GRIMSHAW R. W., ROBERTS A. L. *Gas Research Board Commun.* No. 5, 48 (1946).
61. GRIMSHAW R. W., WESTERMAN A., ROBERTS A. L. *Trans. Brit. Ceram.* Soc. **47**, 269 (1948).
62. GRIMSHAW R. W., WESTERMAN A., ROBERTS A. L. *Trans. Brit. Ceram.* Soc. **44**, 76 (1945).
63. GRIMSHAW R. W., ROBERTS A. L. *Trans. Brit. Ceram. Soc.* **52**, 50 (1953).
64. GRIMSHAW R. W. *Clay Minerals Bull.* **2**, 2 (1953).
65. GRIMSHAW R. W., HARGREAVES J., ROBERTS A. L. *Gas Research Board Commun.* No. 58, 27 (1950).
66. GRUVER R. M. *J. Am. Ceram. Soc.* **33**, 96 (1950).
67. GRUVER R. M. *J. Am. Ceram. Soc.* **33**, 171 (1950).

68. GRUVER R. M. *J. Am. Ceram. Soc.* **34**, 353 (1951).
69. GRUVER R. M. *Am. Mineralogist* **21**, 449 (1936).
69a. HANYKÝŘ V., LEDVINA F. *Silikáty* **5**, 220 (1961).
69b. HANYKÝŘ V. *Sborník VŠCHT Praha* , 383 (1959).
70. HAUL R. A. W., HEYSTEK H. *Am. Mineralogist* **37**, 166 (1952).
71. HENDRICKS S. B., ALEXANDER L. T. *Soil Sci.* **48**, 257 (1939).
72. HENDRICKS S. B., NELSON R. A., ALEXANDER L. T. *J. Am. Chem. Soc.* **62**, 1457 (1940).
72a. HEUMANN T., PREDEL B. *Zeitsch. Elektrochemie* **63**, 988 (1959).
73. HEYSTEK H., SMITH E. R. *Trans. Geol. Soc. S. Africa* **56**, 149 (1953).
74. HONEYBORNE D. B. *Clay Minerals Bull.* **1**, 150 (1951).
75. HOYOS A. DE CASTRO, GONZALES GARCIA F., MARTIN VIVALDI J. L. *An. Soc. Esp. Fís. Quím.* **46B**, 715 (1950).
76. HOWIE T. W., LAKIN J. R. *Trans. Brit. Ceram. Soc.* **46**, 14 (1947).
77. KAUFMAN A. J., DILLING E. D. *Econ. Geol.* **45**, 222 (1950).
78. KEITH M. L., TUTTLE O. *Am. J. Sci,.* Bowen Vol., Pt. 1, 203 (1952).
79. KELLEY K. K., ANDERSON C. T. *Bull. U.S. Bur. Mines* No. 384 (1935).
80. KEATTCH C. J. *Talanta* **14**, 77 (1967).
81. KELLER W. D., WESTCOTT J. F.: *Am. J. Ceram. Soc.* **31**, 100 (1948).
82. KARŠULIN M., TOMIĆ A., LAHODY A.: *I. Rad. Jug. Akad. Znan. Umj.* **276**, 125 (1949).
83. KERR P. F., KULP J. L.: *Am. Mineralogist* **32**, 678 (1947).
84. KERR P. F., KULP J. L., HAMILTON P. K.: *Am. Petrol. Inst. Repr.* No. 3 (1949).
85. KERR P. F., KULP J. L.: *Am. Mineralogist* **33**, 387 (1849).
86. KIYOURA R., SATA T.: *J. Ceram. Assoc. Japan* **58**, 3 (1950).
86a. KLEBER W., FEHLING W.: *Z. Anorg. Allgem. Chem.* **338**, 134 (1965)
87. KULP J. L., WRIGHT H. D., HOLMES R. J. *Am. Mineralogist* **34**, 195 (1949).
88. KULP J. L., TRITES A. F. *Am. Mineralogist* **36**, 23 (1951).
89. KULP J. L., PERFETTI J. N. *Mineral. Mag.* **29**, 239 (1951).
90. KULP J. L., ADLER H. H. *Am. J. Sci.* **248**, 475 (1950).
91. KULP J. L., BROBST A. D. *Econ. Geol.* **49**, 211 (1954).
92. KIEFER C. *Bull. Soc. Franc. Minéral. Crist.* **76**, 63 (1953).
93. KONTA J. *Rev. Central Inst. Geology, Prague* **19**, 137 (1952).
94. KONTA J. *Rev. Central Inst. Geology, Prague* **18**, 601 (1951).
95. KULP J. L., KENT P., KERR P. F. *Am. Mineralogist* **36**, 643 (1951).
96. KRACEK F. C. *Spec. Pap. Geol. Surv. Am.* No. 36 (1942).
97. KRACEK F. C. *J. Phys. Chem.* **33**, 1281 (1929).
98. KURNAKOW N. S., RODE E. J. *Z. Anorg. Allgem. Chem.* **169**, 57 (1928).
98a. LLOYD S. I., MURRAY I. R. *I. Sci. Instruments* **35**, 252 (1958).
99. LEVINSON A. A. *Am. Mineralogist* **40**, 41 (1955).
99a. LEHMANN H., MULLER K. H. *Tonindustrie-Zeitung* **84**, 200 (1960).
100. LIPPMANN F. *Heidelberger Beitr. Mineral. Petrog.* **3**, 219 (1952).
100a. LEHMANN H. *Epitoamyag* **20**, 91 (1968).
101. LONGCHAMBON H., MIGEON G. *Compt. Rend.* **203**, 431 (1936).
101a. MCADIE H. G. *Thermochim. Acta* **1**, 325 (1970); *J. Thermal Anal.* **3**, 79 (1971).
101b. MCADIE H. G. *in Thermal Analysis* Vol. 2 (editors R. F. Schwenker J. and P. D. Garn), p. 1489, Academic Press, New York (1969).
101c. MCADIE H. G. *Temperature Standards for DTA, in Abstracts of Papers of 3. Intern Conf. on Thermal Analysis,* Davos 1971, 1–13.

102. MACKENZIE R. C. *The Differential Thermal Investigation of Clays.* Mineralogical Society, London 1957.

103. MACKENZIE R. C. *Tonindustr. Ztg. Keram. Rundschau* **75**, 334 (1951).

104. MACKENZIE R. C. *Problems of Clay and Laterite Genesis. Am. Inst. Mining Metallugr. Eng.,* New York, 1952.

105. MACKENZIE R. C. *Nature* **164**, 244 (1949).

106. MACKENZIE R. C. *Clay Minerals Bull.* **1**, 115 (1950).

107. MACKENZIE R. C., MILNE A. A. *Mineral. Mag.* **30**, 178 (1953).

108. MACKENZIE R. C., WALKER R. C., HART R. *Mineral. Mag.* **28**, 704 (1949).

109. MACKENZIE R. C. *An. Edafol. Fisiol. Veg.* **13**, 111 (1954).

109a. MACKENZIE R. C. *Talanta* **19**, 1079 (1972).

111. MARTIN VIVALDI J. L., CANO RUIZ J. *An. Edafol. Fisiol. Veg.* **12**, 827 (1953).

112. MCCONNELL D. *Am. Mineralogist* **35**, 166 (1950).

113. MCCONNELL D. *Clay Minerals Bull.* **1**, 179 (1951).

113a. MILLER C. C. *Analyst* **78**, 186 (1953).

113b. MILLER C. C., CHALMERS R. A. *Analyst* **78**, 24 (1953).

114. MOSESMAN M. A., PITZER K. S. *J. Am. Chem. Soc.* **63**, 2348 (1941).

115. MURRAY J. A., FISCHER H. C., SLADE R. W. *Proc. Nat. Lime Ass.* **49**, 95 (1951).

116. MURRAY J. A., FISCHER H. C. *Proc. Am. Soc. Test. Mater.* **21**, 1197 (1951).

117. NAGASAWA K. *J. Earth. Sci. Nagoya Univ.* **1**, 156 (1953).

118. NAGELSCHMIDT G., GORDON R. L., GRIFFIN O. G. *Nature* **169**, 539 (1952).

119. NAGELSCHMIDT G., HICKS D. *Mineral. Mag.* **26**, 297 (1943).

120. NELSON B. W. *The Serpentine-Amesite Join in the System* $MgO—Al_2O_3—H_2O$ and *Classiffcation of the Chlorite Minerals.* Pennsylvania State College, 1955.

121. NORIN R. *Geol. Fören. Stockh. Förh.* **63**, 203 (1941).

122. NUTTIG P. G. *Proof Pap. U. S. Geol. Surv.* No. 197 E, 197 (1943).

123. ORCEL J. *Bull. Soc. Franc. Minéral. Christ.* **52**, 194 (1929).

124. PAGE J. B. *Soil Sci.* **56**, 273 (1943).

124a. PALKIN A. P., PALYURA I. P. *Zh. Neorgan. Khim.* **9**, 2613 (1964).

125. PASK J. A., WARNER M. F. *J. Am. Ceram. Soc.* **37**, 118 (1954).

126. PASK J. A., DAVIES B. *Tech. Pap. Bur. Mines, Washington,* No. 664, 56 (1945).

127. PAVLOVITCH S. *Compt. Rend.* **200**, 71 (1935).

128. POSUJAK E., MERVIN H. E. *Am. J. Sci.* **47**, 311 (1919).

129. PRASAD N. S., PATEL C. C. *J. Indian Inst. Sci.* **36**, 23 (1954).

129a. PREDEL B. *Z. Metallkunde* **55**, 117 (1964).

130. ROBERTSON R. H. S., BRINDLEY G. W., MACKENZIE R. C. *Am. Mineralogist* **39**, 118 (1954).

130a. REMY H., WOLFRUM G., HAASE H. W. *Schweiz. Arch. Angew. Wiss. Techn.* **26**, 5 (1960).

131. ROBREDO J. *Differential Thermal Analysis of Glass. International Commission of Glass,* Charleroi, Belgium 1968.

131a. RACCANELLI A. *Ind. Ital. Cemento* **34**, 3 (1964).

132. RODE E. J., BERG L. G. *Trudy pervogo soveshchanyia po termografii.* Izd. Akad. Nauk SSSR, 1955. Moscow—Leningrad.
RODE E. J. *Iskustvennye soedineniya mineraly i rudy.* Izd. Akad. Nauk SSSR, Moscow, 1952. RODE E. J. *Trudy 4-ogo Soveshchaniya po eksperimental noi mineralogii i petrografii.* Izd. Akad. Nauk SSSR, Moscow, 1951, Vol. 1, p. 95.

133. ROSS C. S., KERR P. F. *Proof Pap. U. S. Geol. Surv. No.* 165E, 151.

134. RIGBY G. R. *Trans. Brit. Ceram. Soc.* **47**, 284 (1948).

135. ROWLAND R. A., LEWIS D. R. *Am. Mineralogist* **36,** 80 (1951).
136. ROWLAND R. A., JONES E. C. *Am. Mineralogist* **34,** 550 (1949).
137. ROY R. *J. Am. Ceram. Soc.* **32,** 202 (1949).
138. SABATIER G. *Bull. Soc. Franc. Minéral. Crist.* **73,** 43 (1950).
139. SASVARI K., HEGEDUS A. J. *Naturwissenschaften* **42,** 254 (1955).
139a. SCHENK M., NACKEN M., KLEESCHULTE H. *Arch. Eisenhüttw.* **31,** 451 (1960).
140. SCHWAB Y. *Analyse Thermique Différentielle.* Publ. No. 22. Centre d'Études et de Recherche de l'Industries des Liants Hydrauliques, Paris 1950.
140a. SCHWIETE H. E., ZIEGLER G. *Z. Anorg. Allgem. Chem.* **B298,** 42 (1959).
141. SEDLETSKII I. D. *Dokl. Akad. Nauk SSSR* **76,** 353 (1949).
141a. SCHOTTMULLER J. C., KING A. J., KANDA F. A. *J. Phys. Chem.* **62,** 1446 (1958).
142. SEDLETSKII I. D. *Zapiski Vesoyuz. Mineral. Obshch.* **78,** 274 (1949).
142a. SELLE J. E., ETTER D. E. *Trans. Metallurg. Soc. AZME* **230,** 1000 (1964).
143. SLADE P. E., JENKINS L. T. *Techniques and Methods of Polymer Evaluation.* Dekker, New York, 1966.
143a. SCHRÄMLI W. *Zement-Kalk-Gips* **16,** 140 (1963).
144. SPEIL S. *Techn. Pap. Bur. Mines, Washington,* No. 66, 1 (1945).
145. SPEIL S., BERKELHAMMER L. H., PASK J. A., DAVIS B. *Techn. Pap. Bur. Mines, Washington,* No. 664, (1945).
146. SPEIL S. *U. S. Bur. Mines. Rept. Invest.* No. 3764 (1944).
147. SCHMIDT E. R., VERMAAS F. H. S. *Am. Mineralogist* **40,** 422 (1955).
147a. SMIDT F. A., MCMASTERS O. D., LIFTENBERG R. R. *J. Less-Common Metals,* **18,** 215 (1969).
147b. SMIRNOV M. P., RUDNICHENKO V. E. *Zh. Neorgan. Khim.* **8,** 1402 (1963).
148. SUDO T. *Science* **113,** (1951).
149. SUDO T. *Clay Minerals Bull.* **2,** 96 (1954).
150. SUDO T., NAKAMURA T. *Am. Mineralogist* **37,** 618 (1952).
151. SUMIN N. G., LASHEVA N. K. *Trudy Mineralog. Muzeya AN SSSR,* No 3, 106 (1951).
152. ŠPLÍCHAL J., ŠKRAMOVSKÝ S., GOLL J. *Věda přírodni* **1B,** 206 (1936).
153. TALIBUDEEN O. *J. Soil Sci.* **3,** 251 (1952).
153a. TAYLOR H. F. W. *The Chemistry of Cements,* Academic Press, Inc., London **1,** 212 (1964).
154. TERTIAN R., PAPÉE D. *Compt. Rend.* **236,** 1565 (1963).
155. THIESSEN P. A., KÖPPEN R. *Z. Anorg. Allgem. Chem.* **189,** 113 (1930).
156. VAN SCHYLENBORGH J., SÄNGER A. M. H. *Rec. Trav. Chim.* **68,** 999 (1949).
156a. VISSER M. G., WALLACE W. H. *Du Pont Thermogram* **3,** (3), 1 (1966).
157. WALKER G. F. *Mineral. Mag.* **28,** 693 (1949).
158. WALKER G. F. *X-ray Identification and Crystal Structures of Clay Minerals.* (Edited by Brindley G. W.) Mineral. Soc. London, 1951, Chapter 7, p. 199.
159. WEISER H. B. *Inorg. Colloid Chem.* **2,** 307 (1935).
160. WEISER H. B., MILLIGAN W. O. *J. Phys. Chem.* **44,** 1081 (1940).
160a. WEFERS K. *Aluminium* **39,** 42 (1943).
161. WEST R. R., SUTTON W. J. *J. Am. Ceram. Soc.* **37,** 221 (1954).
161a. WYLLIE P. J., RAYNOR E. J. *Am. Mineralogist* **50,** 2077 (1965).
162. WITTELS M. *Am. Mineralogist* **36,** 851 (1951).
163. TOPER N. D. *Izv. Akad. Nauk. Ser. Geol.* **29,** 83 (1964).
164. KRIEN G. *Explosivstoffe* **13,** 205 (1965).
165. PLATO C., GLASGOW A. R. *Anal. Chem.* **41,** 330 (1969).

166. RAMACHANDRAN V. S. *Application of DTA in Cement Chemistry,* Chemical Publishing Co., New York, 1968.
167. GARN P. D. *U. S. Clearing House, Fed. Sci. Tech. Inform.,* A D 768 082, 1968.
168. ŠESTÁK J., BERGGREN G. *Chem. Listy* **64,** 695 (1970).

Chapter 5.

APPARATUS

5.1. INSTRUMENTATION FOR TG AND DTA

In the preceding chapters the basic principles of both TG and DTA have been discussed together with the factors which affect the thermoanalytical records and the usefulness of the results obtained. In this chapter the equipment, both home-made and commercially available, will be discussed. The apparatus for DTA and TG will be considered together because in the equipment currently produced both methods are often combined in the one apparatus. Experimental conditions are thus identical in both methods and comparison of the results is more meaningful. However, to a certain extent the conditions will be a compromise and will not correspond to the optimum conditions for the separate methods.

TG requires the following apparatus: (*a*) recording balance with sample holder, (*b*) recorder, (*c*) furnace with programmed heating.

DTA equipment has (*a*) sample holder (often a block), (*b*) furnace with programmed heating, (*c*) recorder, (*d*) detector of thermal potential created in the differential thermocouple, (*e*) amplifier (not always necessary). Some parts are the same for both, notably the furnace with programmed heating, the recorder, and the sample holder. The last-named is one of the most important parts in both systems (see p. 97 ff.).

For a review of the development in the construction of TG and DTA apparatus see Duval [12], Gordon and Campbell [14], Lewin [17], Jacque and co-workers [15], Garn [13], and Murphy [19 − 21].

Laboratory construction of the apparatus is not recommended as considerable experience is necessary, and it is possible to purchase technically advanced apparatus which may easily be modified as required.

The recording balance should have good sensitivity, reproducibility, capacity, insensitivity to changes of ambient temperature, sufficient mechanical and electrical stability, rapid reaction to weight change, and an adjustable range for weight-change recording. Balances may be of deflection or null-point type. Most are null-point, having a built-in sensor indicating a deviation of the balance beam from the zero position, which is restored by means of a mechanical or electrical compensating force. This force, proportional to the weight change, is recorded. Sensory systems include optical systems with a photocell, electronic methods using changes in condenser capacity, or in the inductance of a coil, differential transformers, or change in the intensity of nuclear radiation. Restoring force systems include chains with a servo-mechanism, and electrolytic precipitation or dissolution of an electrode hung on the balance arm. When a deflection-type balance is used the weight change is recorded in a number of forms, e.g. photographically or by recording an electrical signal from a suitable position detector such as an optical or inductive system. The balance itself is usually of beam type, but torsion and spring balances, as well as other systems, have also been used.

The basic requirements of the weighing system are as follows.

1. The balance should be able to record weight changes as a function of time or temperature with sufficient sensitivity and precision.
2. The sensitivity of the balance should correspond to the size of the samples used in the given application.
3. The precision of the recording of the weight change should be better than $\pm 0.01\%$ when all possible disturbing effects, e.g. furnace radiation, convection, magnetic field of the furnace, effect of corrosive gases, etc. are eliminated.
4. The balance construction should be flexible in that it allows for
   (a) the possibility of heating the sample in various dynamic atmospheres or in a vacuum,
   (b) a variable heating rate with the possibility of using an isothermal technique,
   (c) the simultaneous combination of further analytical methods, such as DTA, dilatometry, evolved gas analysis, etc.
5. Easy calibration.

The furnace and the heating system, the additional fundamental parts in both methods, should fulfil the following requirements.

1. The furnace should be usable over a wide temperature range (usually $20-1500$ °C) and permit work in various gaseous media or in a vacuum, with automatic control of the temperature programme.
2. The isothermal zone of the furnace should be sufficiently large for the sample holder to be located within this zone throughout the experiment.

3. The rate of heating must be variable and should be linear and reproducible over the whole temperature range, with a facility for isothermal operation at any desired temperature.
4. The temperature should be measured inside or very close to the sample, with an error of $\pm 1\%$ or less.
5. Conducting and magnetic samples must not interact with the furnace winding.

The method of construction of the furnace and the heating element is usually governed by the required temperature range. Furnaces are described which work at temperatures ranging from very low (-190 °C) to very high (2800 °C). In TG and DTA, furnaces may be used either in a horizontal or a vertical position. In a horizontal furnace a broader isothermal zone may be obtained more easily, whereas a vertical furnace usually has a better axial symmetry and is more suitable for the majority of thermobalances. The furnace may be heated by a resistance heating element, infrared radiation, high-frequency heating, etc. The most commonly used system is resistance winding in the form of a spiral or tape located outside or inside the ceramic tube of the furnace. These windings may be manufactured from various material ("Kanthal", platinum, molybdenum, PtRh, etc.), depending on the required maximum temperature. Some heating materials (Mo, W) require a protective atmosphere or a vacuum. For thermoanalytical curves to be useful, the furnace temperature must increase at a constant rate. The most commonly used heating rates are between 1 and 20 °C/min although in some applications, rates of $0.1 - 300$ °C/min are also used. Any attempt at manual temperature control is bound to fail, and automatic regulation must be used. This may be achieved in various ways. The simplest is to use a variable-voltage autotransformer, the movement of which is controlled by a synchronous motor, regulating the transformer either directly or via a suitable cam. In the case of more complex control systems, a feed-back circuit controlled by a thermocouple in the furnace is used. Control systems are obtainable which fulfil any special requirement of the temperature programme. It should be noted that an "on-off" control system is not suitable, especially in DTA, as it can affect the DTA curve. The temperature in the furnace is monitored by one or two thermocouples connected to one of the control systems already mentioned. With two thermocouples, one is used for the control system and the second for recording the temperature. As was said in the section on temperature measurement, other methods may be used in addition to thermocouples, such as resistance thermometers, thermistors and optical pyrometers.

The recording system should continuously register the temperature, sample weight or in the case of DTA the curve of the ΔT function, or even some function such as the derivative DTA or DTG curve etc. A periodic time record should also be made. The original photographic recording is now replaced by potentiometric recorders of various types. $X_1 - X_2$ recorders are recommended, which are better than $X - Y$ recorders because they permit independent recording of both parameters, In some cases, digital recorders may be required and a feed-out to a computer system has also been used.

The system used to measure temperature difference is a basic part of only the DTA method. Its choice depends on the maximum working temperature, the amplifier and the sensitivity of the recording system, as well as the chemical reactivity of the sample under investigation. The commonest recording system for the DTA curve is differentially connected thermo-couples. The material of which the thermocouples are made depends on the temperature range and the magnitude of the heat of the transformation investigated. Some thermocouples in current use are described in Sections 2.4 and 3.5. If the sensitivity of the recorder is too low for the use of a given thermocouple system, the e.m.f of the differential thermocouple may be amplified by means of a microvolt amplifier. Photographic recording of the DTA curve by means of a mirror galvanometer is rarely used.

The sample holder, including the thermocouple system, is one of the most important parts of the DTA or TG apparatus, and may be common to both of them in combined methods. The effect of the construction of these holders, i.e. the effects of the sample size, crucible shape, position of thermocouples, communication with the atmosphere, etc. were discussed in the relevant sections. The shape and size of the sample holders, as well as the magnitude and the geometry of the sample, have a direct effect on the mechanism of heat transfer, on the reaction, and hence on the recorded curve. In the following section some fundamental types of holders used separately in each of the two methods, and also in the combined method, are discussed along with present possibilities and the further developments.

5.1.1. Sample holders for TG and DTA

Sample holders may be divided into three categories:
(A) Sample holders for DTA.
(B) Sample holders for TG.
(C) Sample holders for combined TG and DTA.

A. Sample holders for DTA.

These are constructed as massive ceramic or metallic blocks provided with external or internal heating, and possibly with special thermocouple probes (see Fig. 5.1).

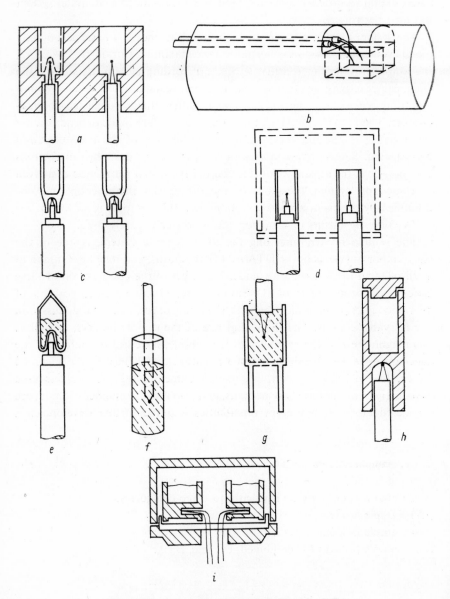

Fig. 5.1 Various types of sample holders for DTA.

1. The arrangement of the DTA block for a vertical (Fig. 5.1a) or horizontal furnace (Fig. 5.1b) allows location of the thermocouple junctions either directly in the sample or in the crucibles. In the vertical arrangement a larger number of samples can be simultaneously analysed by having a larger number of cavities in the block [16]. Access of the atmosphere to the sample is limited in this case.

2. The arrangement of the sample holder for use in a vertical furnace, which does not use a massive block but has separated thermocouple probes, usually has round-bottomed crucibles or crucibles which can be attached to the thermocouple junction (Fig. 5.1c). The use of metallic or ceramic tubes and the location of the samples in direct contact with the thermocouple junction (Fig. 5.1d) is also possible. Such arrangements are found in the majority of commercially produced equipment for DTA. In such cases, probes may also be smaller, and are used with micro-crucibles or tubes for only several tens of milligrams, with a protective cover outside, similar to the arrangement for dynamic differential calorimetry (see Fig. 3.28).

3. Volatile or corrosive samples may be placed in a special vessel made of quartz, glass, etc., which may either be sealed or closed with a special lid (Fig. 5.1e). If the thermocouple may make contact with the substance investigated (even a liquid), the arrangement with test-tubes shown in Fig. 5.1f may be used. This arrangement may also be used with thin-walled tubes protecting the thermocouple. The access of gas to the sample may be improved by using a crucible with a porous bottom through which gas may pass (Fig. 5.1g). For isobaric measurement special crucibles, usually of metal, are used, having tightly fitting lids (Fig. 5.1h). The use of special metal crucibles, isolated from the external metal jacket, is shown in Fig. 5.1i (see Section 3.3) [11].

4. A number of other thermocouple probes have been described, with special applications. Mazières [18] used chromel-constantan thermocouples, the junctions of which were shaped in the form of dishes in which only a very small amount of sample (solid or liquid) could be placed. The sensitivity of this arrangement is sufficient for micro-determinations (see Section 3.9.6, Fig. 3.27). There are also thermocouple junctions made by welding Pt and PtRh tubes, to form a crucible for the sample. Pakulak and Leonard [23] described a special test-tube arrangement using as detectors thermistors located symmetrically in glass vessels. Numerous other special arrangements could be mentioned which permit simultaneous analysis of gaseous products, work at high or low pressure, work in a vacuum, etc.

B. Sample holders for TG

The development of thermobalance construction and various types of sample holders is discussed by Duval [12]. The construction of the sample holder depends on the construction of the thermobalance, and on whether the sample is located above, below or to one side of the balance. For a sample hung below the balance the simplest method is a hanging crucible. This arrangement, however, makes measurement of the sample temperature difficult. A possible arrangement is shown in Fig. 5.2a, in which the

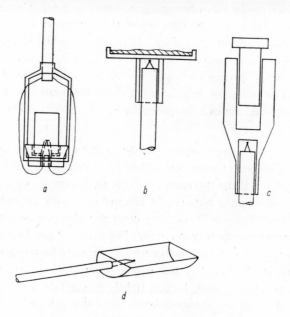

Fig. 5.2 Various types of sample holders for TG.

thermocouple junction is introduced into the concave bottom of the crucible Placing the sample above the balance makes the measurement of the temperature of the sample or of the crucible bottom simple (Fig. 5.2b, c). At the same time, this arrangement permits performance of experiments under conditions of constant pressure or with access of the furnace atmosphere. The arrangement in Fig. 5.2 also allows the use of the type of crucible used in the apparatus known as a Derivatograph [24]. The horizontal arrangement requires construction of the carrier in the form of a small boat (Fig. 5.2d). Measurement of the sample temperature is rather difficult under these circumstances.

C. Sample holders for combined TG and DTA

Sample carriers for simultaneous TG and DTA have been developed which are both simple and versatile. Among the many types which have been developed, several basic modifications which are used commercially will be dealt with here.

A simple arrangement in which the reference thermocouple is located in the hollow of the probe is shown in Fig. 5.3a [2]. The arrangements shown in Fig. 5.3b, c, are those commonly used today (Netzsch, Linseis, etc.); they can be provided with a more massive ceramic or metallic block (TEGRA II, Stanton), or with a completely separate crucible with the reference sample placed outside the balance suspension (Derivatograph). The construction shown in Fig. 5.3d is suitable for horizontal arrangement of the apparatus. The use of differentially connected boats which serve simultaneously as thermocouple junctions is more difficult. The arrangements shown in Fig. 5.3e, f, g, h are those of the Mettler balances. In Fig. 5.3e the arrangement is for microanalytical methods; f and g permit the temperature field to be made uniform; in g the external jacket is also the thermocouple junction. The arrangement shown in h permits simultaneous use of three crucibles.

5.2 COMMERCIAL APPARATUS FOR TG AND DTA

Both thermoanalytical methods have developed rapidly over the past twenty years, largely owing to the improvement in commercial equipment for TG and DTA either separately or combined. A number of excellent designs are available, with high standards of sensitivity and precision. They are usually very versatile and may be used at high temperatures. The choice of apparatus should take into account factors such as sensitivity, precision, temperature range, reaction atmosphere, vacuum facilities, sample weight, etc. Individual requirements may be met in some cases by adjustment of the sample carrier and the reaction space, preserving all the original basic constructional parts.

To facilitate the choice of apparatus, technical parameters of the best known apparatus are given here. In some firms (mainly German, British, French, and American) manufacture of this apparatus has a long tradition of reliability and a variety of types is produced. Data and technical parameters are taken mainly from advertising literature. The exclusion of any manufacturer is unintentional and implies no adverse criticism of the equipment.

In Czechoslovakia a number of papers have appeared dealing not only with the application of the method of thermal analysis and with its theoretical aspects, but also with practical solutions of constructional problems. The author of this book and his co-workers constructed, between 1957 and 1961, four types of thermobalance with direct electronic recording, allowing simultaneous TG and DTA in a vacuum or a controlled atmosphere, as well as an independent arrangement for DTA [1 – 8].

Another apparatus constructed in Czechoslovakia is the thermobalance manufactured by the Research Institute of Radiocommunication (TESLA), and based on the ideas of Beránek [9, 10]. It is a device for thermal analysis and for thermal treatment, constructed so that various types of furnaces with detectors for various types of measurements may be connected with the basic controlling and recording unit. The furnaces are modular and

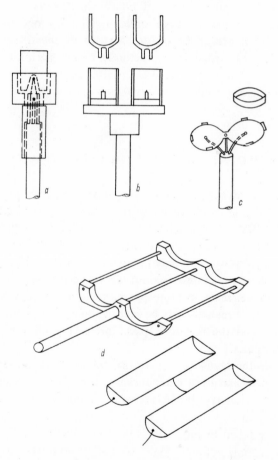

Fig. 5.3 Various types of sample holders for combined TG + DTA.

usable in the vertical or the horizontal position. "Kanthal" or "Crusilite" is used as the heating material. For DTA, vertical furnaces with massive ceramic blocks are used, adapted for round-bottomed crucibles into which thermocouples protrude. For TG, electronic compensation thermobalances

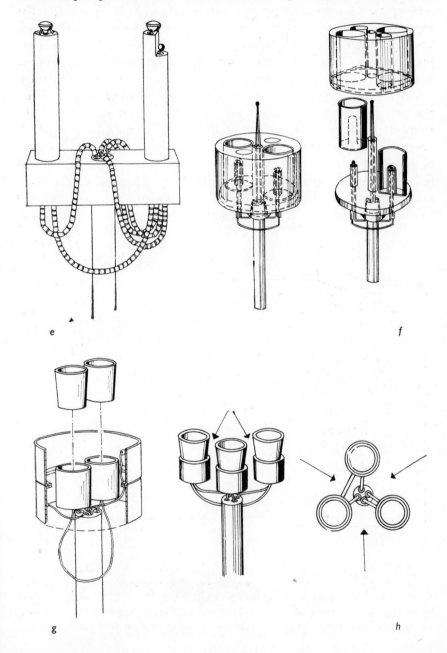

e

f

g

h

located below the furnace are used. The balance has the centre of gravity shifted to the axis of rotation of the balance beam, permitting complete compensation of the weight change of the sample. The compensation is electromagnetic, controlled by a semiconductor amplifier regulated by a photoresistance operated by an optical lever on the balance beam. The weight change can be recorded or read directly. The device may be fitted with an automated furnace for ignition in a temperature gradient and the preparation of hardened samples for microstructural X-ray analysis, as well as with a dilatometer. The controlling recording unit contains a current source for the furnaces with a compensation temperature regulator, a temperature programmer, a device for compensation measurement of the temperature and a compensation recorder. The compensation regulator operates by adjusting the heating current to the optimum value, and by causing small oscillations of this current that are too rapid for the furnace to follow, so that the oscillation of the temperature is minimized. The maximum temperature of the furnace is 1550 °C. The apparatus is manufactured to order, according to the special requirements of the customer.

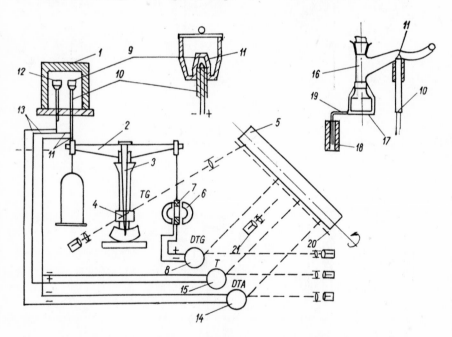

Fig. 5.4 Schematic diagram of the Derivatograph: *1* — furnace; *2* — balance beam; *3* — balance pointer; *4* — mirror; *5* — recording drum; *6* — permanent magnet; *7* — coil; *8, 14, 15* — galvanometers; *9, 12* — crucibles; *10* — sample holder; *11, 13* — thermocouple leads; *16* — vessel for distillation; *17* — heated carrier; *18, 19* — heating; *20* — copying stencil of the temperature and weight scale; *21* — lamp for template copying.

In Hungary an apparatus known as the Derivatograph, type OD-102, is constructed according to the original papers of Paulik, Paulik and Erdey [24 – 27]. This arrangement gives simultaneous TG, DTG, and DTA measurements. Its constructional simplicity is mainly due to the fact that the balance functions on a deflection principle and the recording of all investigated quantities is photographic. The general scheme is shown in Fig. 5.4. The sample holder (10) is placed above the balance beam, in the resistance furnace in which the probe with the reference material (12) is also located; the reference material is not weighed. The weight change is recorded on photographic paper by means of a light-beam reflected from a mirror (4) on the balance pointer. By means of a special coil (7) suspended on the balance beam and moving in the magnetic field of a permanent magnet (6) the rate of weight change, which is proportional to the current induced in this coil, is measured simultaneously, and is recorded via the galvanometer (8) (DGT). The DTA curve is also recorded photographically, via a sensitive galvanometer (14). The same holds for recording of the temperature of the bottom of the crucible containing the sample. The temperature and weight calibration scale may be photographically copied on to the recording paper by means of a template. The sensitivity of the balance is ±0.4 mg and full scale deflection covers the range 20 – 2000 mg in seven steps. The galvanometers have calibrated shunts. The photographic paper is wound onto a drum rotating at the required rate, 50, 100, 200 or 400 min/rotation. The furnace is doubly wound allowing linear heating up to 1000 °C at rates between 1 and 20 °C/min, by means of a cam-type temperature programmer. As supplied, the apparatus will not function in a controlled gaseous medium. A number of papers have been published which describe adjustments of the sample carrier and the reaction space to permit use of this arrangement for special purposes, e.g. work in controlled atmosphere, simultaneous evolved gas or dilatometric analysis [28, 29]. The disadvantages of this method are the change of the position of the sample relative to the isothermal zone of the furnace during the experiment, and the photographic recording, which is not usual nowadays.

In West Germany several types of thermobalance are manufactured, primarily those made by Sartorius, Linseis, and Netzsch, all of which have been developing apparatus for TG and DTA for a number of years.

Sartorius-Werke Göttingen currently make two basic sets of apparatus, an electronic microbalance and an ultramicrobalance (series 4100) and also a universal measuring device, Gravimet (series 4300). The series 4100 microbalance, the construction of which is shown in Fig. 5.5, is based on the principle of independent feed-back compensation of the balance beam deflection; the balance beam is constructed from a quartz tube (5) (because of its low thermal expansion) connected to a coil (1) which is placed in the air space of a system of permanent magnets. When current is passing, a deflection proportional to the current intensity is observed. Parallel to one side of the coil two additional coils (4) are fixed on a carrier (3). The coils carry, in opposite directions, a high-frequency alternating current. In the zero position of the balance beam potentials induced in the coil (1) are cancelled. However, when the balance beam is deflected an alternating current is induced in the coil (1), the amplitude and the phase of which depend on the magnitude and the direction of the deflection. This potential induces the regulation potential which feeds the regulating coil with direct current, causing compensation of the deflection. The coil with the balance beam is fixed on tensioned metallic strips (2).

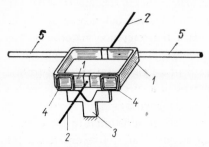

Fig. 5.5 Functional scheme of Sartorius microbalance.

1 — coil; *2* — axis of the torsion wire suspension; *3* — fixed carriers; *4* — coil of the alternating current field; *5* — balance beam.

Nine different types of balance are made which differ according to the construction of the balance beam and coil system. Models 4102 and 4105 are vacuum microbalances for up to 5 g sample weight. Models 4104 and 4107 are of the same type, for samples up to 50 g. Model 4112 is a high-pressure microbalance for up to 5 g load. Models 4142, 4144, 4145, and 4147 are the basic balance systems of the preceding type, which can be provided with a special vacuum device. The technical parameters of these thermobalances satisfy the most exacting requirements and cover the whole field of required sensitivities and measuring ranges. For example, model 4102 covers the range from 2 to 200 mg, and model 4104 the range from 20 mg to 2 g. It is also possible to work at high vacuum (10^{-10} mmHg) in some models (4144, 4145, and 4147). The balance may be connected to an electronic recorder or a digital voltmeter. All the models mentioned can be provided with numerous accessories, such as quartz reaction jackets and sample suspensions, tube furnaces for temperatures up to 600 °C, and programmed temperature control.

Another Sartorius product is a universal gravimetric measuring device, Gravimet, type 4303. The apparatus records simultaneously the weight change, the amount of the gas absorbed by the solid substance, the temperature, and the pressure. The apparatus is suitable for measuring adsorption, and desorption of gases and vapours by solid substances as a function of pressure (10^{-6} — 800 mmHg) and temperature (from

—200 °C to +500 °C). In a certain respect this apparatus represents a
special application of the thermogravimetric method for the following of
sorption isotherms and of the kinetics of sorption processes.

In collaboration with the firm of Heraeus, a high-temperature thermobalance has been developed making use of the Sartorius microbalance
model 4101 as the weighing system. (An older type, the precursor of model
4102, was based on the same principle.) This thermobalance is provided
with an electrical resistance furnace. The high-temperature zone is
concentrated in a small space and enables the sample to be heated to
1600 °C. The construction of the furnace and of the regulating system
permits maximum temperature to be attained in 5 minutes. The sample may
also be heated or cooled at a chosen rate. The sample is placed in a
platinum-metal dish suspended below the balance. If the furnace section
is duplicated the apparatus may be used as a derivative thermobalance in
view of the symmetrical construction of the weighing system. The advantage
of the feed-back system used in all the types of Sartorius balances
mentioned lies in the unchanged position of the sample relative to the
isothermal zone of the furnace.

In 1970 Sartorius put on the market a new type of thermobalance called
Thermomat [33]. This apparatus used an original design for the connection
of the sample suspension to the balance beam. A special magnetic coupling
is used, giving complete separation of the balance beam space from the
reaction space.

The Netzsch firm (Selb/Bayern, West Germany) has developed a series
of apparatus for TG and DTA. The current range includes a thermobalance
for a controlled gas atmosphere, model 409, a vacuum thermobalance,
model 419, a device for DTA, and complete equipment, including amplifiers, furnaces, and control units for DTA.

The automatic balance, model 409, is a classical type of thermobalance
with a sensitivity of 0.3 mg with a sample carrier located above the balance.
Its weighing system consists of a three-knife balance beam on one end of
which the suspension with the sample holder is fixed, while the other end is
provided with a chain for compensating the beam deflection. The balance
operates by a compensating system, the weight change being compensated
continually by a regulating chain controlled by a servo-motor, in turn
controlled by a photoelectric system. This arrangement automatically
records weight changes up to 100 mg, at an effective sensitivity of ±0.5 mg.
In more recent designs, electromagnetic compensation of the balance beam
deflection is used; the beam is composed of a coil, a permanent magnet,
movement recorder, and an amplifier. The advantage of the system used is
the unchanged position of the sample in the furnace. The furnace itself is

made of a tube of silicon carbide, 40 mm internal diameter, with maximum temperature 1550 °C. In order to limit heat radiation and convective flow of the atmosphere special shielding is used. For temperature regulation a regulator (model 406) is used provided with a thyristor regulator. It controls the temperature of heating and cooling at ten rates in the range from 0.1 to 100 °C/min. It also allows automatic switching off after the required temparature is reached, maintenance of isothermal conditions, and automatic switching to programmed cooling. The temperature programme is preselected with the aid of a template. The temperature is measured in close proximity to the sample crucible.

Another product of the Netzsch firm is a vacuum thermobalance, type 419. It works on the same principle as the balance described above, the only difference being that the balance casing and furnace are constructed in a manner allowing evacuation down to 5×10^{-5} mmHg. The thermobalance comes complete with an evacuation system and a gauge, and an $X-Y$ recorder for the recording of weight changes as a function of temperature, or a point compensation recorder.

Another product of this firm is an apparatus for DTA, model 404/406. Furnace and temperature regulation units are the same as in the thermobalances. Netzsch manufactures a series of sample holders and measuring heads and also equipment for dynamic differential calorimetry (DDC, see Fig. 3.28). This choice of measuring heads permits work in a number of different atmospheres or in a vacuum, with a range of sensitivities and resolving powers, various amounts of sample (either solids or liquids), up to temperatures of 1550 °C with a maximum sensitivity of 0.01 °C/mm. A special measuring head for low temperatures allows work in the range from —160° to 420 °C with 0.005 °C/mm sensitivity.

Another West German firm, Linseis K.G. Selb/Bayern has developed a series of thermobalances and apparatus for DTA. At the moment it offers thermobalances of types L 72, L 81 and L 82, a microbalance, and a magnetic balance L 86. This last is an automatic recording balance employing electromagnetic compensation. A change in position of the balance beam is indicated by means of a bridge connection fed by an oscillator of 10 kHz frequency. The alternating voltage appearing across the bridge is proportional to the angular displacement of the balance beam and also indicates the sign of the weight change by its phase. The balance beam is suspended on a tensioned strip of beryllium bronze. The construction of the balance is interesting; it may be used in a vertical position, with the sample suspended from it, or placed below the furnace, as well as in a horizontal arrangement. The balance housing, including the reaction unit, is made so as to permit work in a controlled atmosphere, or in a vacuum down to 10^{-5} mmHg. The nominal balance sensitivity is 10^{-5} g within ranges from ± 10 to ± 300 mg, with the possibility of increasing the weighing ranges to 3 g when use is made of compensation weighing. Maximum weight of the sample is 5 g. The sample carrier construction allows simultaneous recording of the DTA curve, both in the horizontal and vertical arrangements of the balance. The apparatus may be obtained in combinations TG + DTA, or TG, DTG + DTA. For the latter purpose

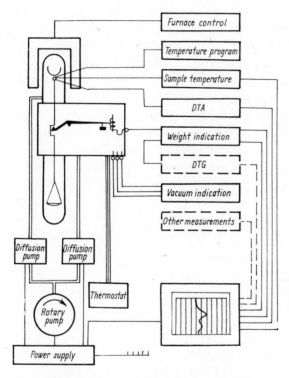

Fig. 5.6 Block diagram of Mettler thermoanalyser.

it is also fitted with a special three-pen recorder. The furnace may be one of three types ("Kanthal", graphite, SiC) covering the temperature range $20 - 2200$ °C. The temperature programme regulator uses thyristors and gives linear heating at rates of $0.1 - 50$ °C/min or cooling. The model for low-temperature measurements covers the range from —200 to +500 °C. The regulator functions on the compensation principle, by comparing the e.m.f, of the furnace thermocouple with the programmed potentiometric value. The difference in these values is amplified and monitored by the impulse generator controlling the thyristors.

Another product of Linseis is an apparatus for DTA. The furnace, programming and recording parts are the same as the components of the thermobalance. The measuring head, with differentially connected thermo-couples, is used in the vertical position. The device is supplied with various exchangeable parts, allowing direct contact of the thermocouple junction with the sample, use of special crucibles, or microscale operation (see Fig. 3.27).

The Swiss firm Mettler, Zürich, manufactures a vacuum recording thermoanalyser, consisting of a thermobalance combined with a simulta-neous recording of DTA or other curves, such as DTG curves, pressure curves, etc. A block diagram of the apparatus is shown in Fig. 5.6. The basis of the whole arrangement is an unequal-arm automatic recording balance, working on the compensation principle. The change in balance beam po-sition is indicated photoelectrically and the compensation is made electro-magnetically. The current passing through the coil of the electromagnet is proportional to the force necessary for the compensation; it causes a po-tential drop across a calibration resistance and this represents the value of the weight change. The apparatus records two weight-change curves simulta-neously, one the overall curve and the other for accurate evaluation, the latter being divided into regular intervals of constant weight change. The balance has three ranges, both for automatic compensation and for the measuring ranges of the recorder. For the recorder curve the ranges are:

Measurement	I	II	III
Overall curve	$0-1000$ mg	$0-100$ mg	$0-10$ mg
Detailed curve intervals	100 mg	10 mg	1 mg

This arrangement satisfies even the most exacting requirements, as at the most sensitive range a weight change of 5 µg may be recorded. The apparatus is constructed so that the sample holder is above the balance and is provided with shielding against radiation and convection currents. Sample carriers are easily exchanged and allow simultaneous recording of DTA curves. The measuring heads for combined analysis are very carefully made and some of them are shown in Fig. 5.3. Generally, the reference material is weighed together with the analysed sample. For TG only, various types of crucibles are supplied, made of PtRh, Al, or Al_2O_3. Various measuring heads used in conjunction with these crucibles allow macro and microanalysis; the sensitivity ranges mentioned are adapted to this. Temperature detec-tion is by thermocouple junctions in contact with the crucible bottom, or located in a special cavity in the centre of the crucible bottom. The construction of the reaction unit and of the balance beam jacket, which is double and fitted with a thermostat, as well as of the retort, permits measurement in a static or dynamic atmosphere with inert or corrosive gases, or work in a vacuum down to 8×10^{-6} mmHg. The amplifier for the DTA curve has six sensitivity ranges, from 20 to 1000 μV, and is fitted with a continuous base line compensation at all ranges. The TG curve can be differentiated electronically to give the DTG curve. The apparatus may be equipped with several types of furnace covering the range 25–1600 °C, as follows.

(a) Low-temperature furnace for temperatures up to 1000 °C. This is a vacuum-tight furnace with a "Kanthal" twin spiral located in a quartz jacket.

(b) High-temperature furnace for temperatures between 25 and 1600 °C. This is a vacuum-tight "Superkanthal" furnace with a corundum reaction unit,

(c) Furnace for special atmospheres containing water vapour. This is a furnace in a quartz jacket, with a twin "Kanthal" spiral for temperatures up to 900 °C. The furnace jacket is connected to an evaporator and condenser. The programmer permits the choice of various temperature programmes, as well as linear heating or cooling at ten selectable ranges, from 0.5 to 25 °C/min. The system is fitted with various improvements, such as fivefold amplification of the temperature curve (for more precise readings of temperature), a device for sampling the reaction gases, an auxiliary device for linear heating in the 28–80 °C interval, facility for connecting an additional measuring system, e.g. evolved gas analyser, and other auxiliary devices, such as a vibrating spatula for filling the crucibles, etc. The quality (and price) of the apparatus is high, and its accuracy, sensitivity, and design enable quantitative DTA measurements to be made.

In France, the first country to produce a thermobalance commercially, several types of thermobalance are available. The history of the development of the first commercial thermobalance, due to Professor Chevenard, is described by Duval [12]. The producer of this apparatus was the A.D.A.M.E.L. firm in Paris, today producing a vacuum thermobalance based on the original Chevenard construction.

The company SETARAM (Société d'études d'automatisation de regulation, et d'appareils de mesures), Lyon, offers two types of thermobalance. Thermobalance B 60, type Ugine-Eyraud, has been made for a number of years, and has been modified from time to time. This is an automatic recording balance with a classical equal-arm beam, suitable for work in a controlled atmosphere at various pressures or in a vacuum. The internal arrangement of the weighing system is shown in Fig. 5.7. The balance functions on the compensation principle, using electromagnetic resetting of the beam combined with a photocell position recorder. The intensity of the light-beam directed toward the photocell varies as a function of the balance beam deflection. The current passing through the photocell is amplified and passed through a coil in the axis of which a permanent magnet moves, suspended on the arm of the beam. In addition to this automatic compensation system, the balance is fitted with automatic addition of weights serving for weighing the sample and calibration during the experiment. The maximum load is 100 g with a sensitivity of 0.1 mg. The automatic compensation weighing can be done in 10 ranges, from 0 to 3.2 g. The balance jacket is constructed for work in a vacuum (down to 5×10^{-6} mmHg) and at pressures up to 200 g/cm^2. The sample to be analysed is suspended below the balance in a sample holder which permits the location of three crucibles and a simultaneous recording of DTA, TG, and DTG curves. Corundum or metal (PtRh or tantalum) crucibles are employed which are placed directly on the thermocouple junctions. The measuring head of the suspension is made of corundum

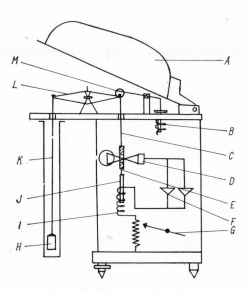

Fig. 5.7 Functional scheme of Ugine-Eyraud thermobalance B 60.

A — balance-beam jacket; B — electromagnet for tare loading; C — suspension; D — photocell; E — aperture; F — amplifiiers; G — recording system; H — crucible with sample; I — coil; J — magnetic core; K — sample suspension; L — balance beam; M — tare load.

or quartz. The furnaces are of three types ("Kanthal", graphite, tantalum), covering the range 20 − 1600 °C. The temperature programmer gives linear heating over a wide range of rates, or permits work under isothermal conditions. The whole apparatus is constructed so that it permits a simultaneous analysis of gaseous products. The SETARAM company modified the gas chromatograph Chromodam in such a way that it could be connected to the thermobalance B 60.

Another product of the SETARAM company is the thermobalance type Electrodyn MTB 10-8. In principle it is an equal-arm microbalance, the beam of which is suspended from a torsion wire and the deflection compensated electromagnetically; there is photo-optical recording of the movement of the beam and the solenoid. The apparatus has a maximum loading of 10 g with 4×10^{-8} g sensitivity. The lowest detectable weight change is 0.4 µg, corresponding to 1 mm on the record. Nine sensitivity ranges cover weight changes up to 40 mg. Automatic electromagnetic weighing (in the $0-250$ mg range) permits calibration and the alteration of the ranges during the weighing. The apparatus is capable of working under vacuum down to 10^{-6} mmHg or in a controlled atmosphere. Symmetrical suspension of sample from both ends of the equal-arm beam permits differential weighing. For this purpose a double symmetrical furnace is constructed, with maximum temperature 1000 °C. The temperature regulator, type PRT 3000, allows a choice of linear heating rates at up to 25 °C/min. A simultaneous recording of the DTG curve may be made.

The SETARAM company also manufactures an independent unit for DTA. The apparatus is available in two modifications, for temperatures up to 1000 °C or 1600 °C. Temperature regulators and the recording system are the same as in thermobalance B 60. The apparatus is fitted with three interchangeable heads, a standard model for temperatures up to 1000 °C, the other two for work in a vacuum or on the microscale. The thermocouples are PtRh and AuPd. The measuring head is situated in a quartz chamber and the construction permits work in controlled atmospheres. The recording system simultaneously records the DTA curve and the temperature.

The firm Stanton of Great Britain currently produces three basic types of thermobalance, two employing macro samples, and the most recent operating on the micro scale. The standard macro model operates in air, and the "Massflow" operates on the macro scale with a controlled atmosphere (vacuum down to 10^{-4} mmHg or gases, even corrosive ones, at various pressures). Both types have a constant-load weighing-beam system, and in the case of the vacuum type the twin balance beams are magnetically coupled. The system works automatically, making use of a servo-mechanism and allowing automatic weight loading. In this system several fundamental types of furnaces and constructions of weighing parts are combined, leading to a wide choice of models and permitting the use of this apparatus in most fields. Combining this thermobalance with an independent apparatus for DTA with sensitivity ranges of $50-500$ µV allows TG and DTA to be performed simultaneously. The sample holder is located above the balance in all models and in the case of the combined method is terminated by a ceramic block with cavities for the crucibles for the sample and the standard, which are weighed together. The third type of balance operates from room temperature to 1000 °C and uses a low-mass water-cooled furnace in conjunction with an electronic microbalance. In addition to the thermobalance models mentioned four types of independent recording balances are also produced, differing in sensitivity and maximum load, and having different ranges of automatic weight loading.

American and Japanese firms produce wide ranges of apparatus. In view of the number of instruments offered, especially with respect to the automatic recording balances, only apparatus which appears to have broad application will be described here.

The firm Perkin-Elmer, Norwalk, Connecticut, U.S.A. introduced onto the market a system for thermal analysis composed of a TGS-1 thermobalance and a DSC-1 B differential calorimeter. This system represents an important step in the transition to miniaturization of the methods of thermal analysis, which affords a series of advantages discussed earlier. The thermobalance TGS-1 is based on the Cahn RG electrobalance which can work with very small samples and eliminates the complications of thermal and pneumatic interferences encountered in many systems. This weighing system functions on a compensation principle. A photocell system detects the deflection of the balance beam, and an electromagnetic system, similar to that in the Sartorius microbalance, compensates for the deflection. The balance beam is fitted with three suspensions, the first for sample weights in the range 0 – 200 mg, the second for the range 5 – 1000 mg, and the third for compensation of the sample weight. The whole balance beam system is enclosed in a glass jacket, permitting work in a static or dynamic gas atmosphere, or in a vacuum. An interesting recent development of this arrangement is a miniature furnace of very low weight located inside the down tube. This permits very rapid controlled heating and cooling of the samples without any disturbing aerodynamic effects. The furnace is fixed from below into the reaction unit at the site of the first or the second balance beam suspension. In addition to the advantages following from the general miniaturization of the method it also has an original method of temperature calibration based on reversible magnetic transitions in ferromagnetic substances. The dimensions of the furnace allow the location of a permanent magnet in close proximity to the sample; the magnetic force of the ferromagnetic substances affects the balance beam as a mass would do, and the changes of magnetic properties simulate the weight changes. At the temperature of the Curie point a sharp wave appears on the TG curve, corresponding to the characteristic temperature (see Section 2.4). This temperature is independent of the atmosphere and pressure and corresponds to the temperature inside the sample. In a single experiment a number of characteristic temperature points may be obtained and reproduced, if necessary, several times with the same sample. The TG and DTG curves may be recorded simultaneously. The miniature resistance furnace allows the heating of the sample in the temperature range from 20 to 1000 °C, and the temperature programmer UU-1 allows heating or cooling at eleven linear rates, ranging from 0.31 to 320 °C/min.

A further part of this system is the DSC-1B calorimeter operating at a different temperature interval, i.e. from −100 °C to +500 °C. The principle of the DSC method was explained in the section on special methods of DTA (3.9). The principle of this method is shown in Fig. 3.29. The device is composed of three units, viz. analyser, control unit, and recorder. The analyser is separable which permits its location in the furnace or elsewhere. It comprises a sample carrier, including a device for its heating, and the equipment necessary for temperature calibration. The control cabinet contains all the electronic parts of the system. The temperature programmer permits linear heating or cooling. It has eight ranges, from 0.5 to 70 °C/min. The sensitivity of the detection may be chosen in a range of 1 − 32 mcal/sec (for full scale deflection). The accessories of the apparatus are numerous and comprise an adaptor for very low temperature, special crucibles, sapphire standards, and calibration lists. The apparatus is of high quality and points to new developments in thermal analysis techniques.

Similar equipment for combined thermoanalytical techniques is manufactured by Du Pont Instruments, Wilmington, Delaware, U.S.A. This firm has also introduced onto the market a thermal analysis system composed of a DSC calorimeter and a thermobalance, and also classical DTA ranging up to 1600 °C. The equipment is available in modular form. The basic module comprises the recorder, temperature programmer, electronics, and the gas-flow controller, which are arranged side by side in a panel. Numerous accessories are connected directly on the top of the basic module, or onto the panel. These consist of the detector of the DSC calorimeter and the detector for DTA (four different types covering the temperature range between −100 °C and +1600 °C). The thermobalance and an independent thermomechanical analyser are located in separate modules. The thermochemical analyser determines mechanical deformations as a function of temperature by means of a linear differential transformer. The main module is fitted with an $X-Y$ recorder and a temperature programmer devised for a broad choice of heating and cooling rates and isothermal working conditions. The principle and the parameters of the DSC module are in principle identical with the Perkin-Elmer type. It should be noted that a similar DSC system is also manufactured by the Japanese firm Rigaku Denki Co., the third manufacturer of this type of apparatus. The DTA system is composed of four types of recording systems: a standard type for temperatures up to 500 °C, a type for temperatures up to

850 °C, and high-temperature models for temperatures up to 1200 °C or 1600 °C. The last two are of the calorimetric type and use the measuring heads shown in Fig. 5.1. The thermobalance type 950 is in an independent unit which may be connected to the basic module. It is a null-point semi-microbalance with electromagnetic compensation, a capacity of 300 mg, and seven ranges from 1 to 100 mg with the possibility of automatic weight compensation up to 110 mg. The temperature programme covers a range of heating rates 0.5 – 30 °C/min up to maximum temperature 1200 °C. The balance can work in a vacuum and in controlled atmosphere.

Another American product is the thermoanalytical system of Fisher Instruments Manufacturing Div., Indiana, consisting of a thermobalance model 100 and an independent DTA unit. This thermobalance also makes use of the Cahn RG model electrobalance (similar to Perkin-Elmer), with the sample suspended below the balance beam, but is fitted with the classical resistance furnace outside the balance casing, which itself permits work in a vacuum or a controlled atmosphere. The balance is designed for a maximum load of 2500 mg, at a sensitivity of 0.1 µg. The largest weight change that can be recorded is 1000 mg. The furnace has a maximum temperature of 1200 °C and eight heating rates, ranging from 0.5 to 25 °C/min. An accessory programmer allows the extension of the heating rate from 0.05 to 150 °C/min. For a more accurate reading of temperature the apparatus is provided with a device for partial suppression of the e.m.f. of the thermocouple. The TG and DTG curves are recorded, but when the apparatus is combined with a separate block with crucibles for DTA and with an analyser 260, a DTA curve may also be obtained. This curve is not recorded from the weighed sample, but from an independent DTA block. The arrangement is unusual, because both thermocouple junctions are inserted from above. The block is provided with cavities into which two pairs of quartz crucibles are inserted, while two pairs of differential thermocouples are inserted into the samples from above.

In Japan several interesting sets of apparatus for TG and DTA are manufactured. Some of the apparatus has been miniaturized, e.g. the products of Shimadzu Seisakusho Ltd., Kyoto. The thermogravimetric apparatus type DT 20 B 2 R-2022TFGA-20 is composed of a torsion micro-balance based on the compensation principle, using feed-back electromagnetic compensation and photocell indication of the balance beam position. The device has a 1 – 200 mg range of recorded weight changes and has high sensitivity and reproducibility. This firm also produces micro-DTA apparatus, type DT 20 B + R 202 + MDM 20, working in the temperature range 20 – 600 °C. In addition, the Shimadzu firm produces a series of devices for classical TG and DTA. The thermobalances of type TB-2 B and type

DTG-2 B are normal beam balances working on the compensation principle, a solenoid and a differential transformer in a feed-back circuit. These instruments have a sensitivity of 0.25 mg at five selectable recording ranges from 50 to 800 mg for full deflection of the recorder and can be used with a vacuum or a controlled atmosphere, up to a maximum temperature of 1200 °C. The DTA instrument, type DT-2 B, is a classsical type of apparatus with a vertical arrangement and a robust block suitable for performing a series of analyses. A similar instrument of classical type, but with a horizontal arrangement, is produced by Mitamura Riken Kogyo Inc. (model 1702 DR).

The Japanese firm Rigaku-Denki Co., Tokyo, produces instruments for TG, DTA and calorimetry. In addition to automatic recording thermo-balances working on the compensation principle and simultaneously recording TG, DTG, and DTA curves, they also produce a separate apparatus for DTA. This is the so-called DTA or SHM unit. This instrument is interesting in that it is fitted with a spherical furnace with internal resistance heating, which may be used in three different ways by suitably adjusting the recording part, as an apparatus for DTA, as an adiabatic calorimeter, and as a differential calorimeter. The adiabatic method of Nagasaki and Takagi [22] is used for heating. The difference between the temperatures of the cover and the block is recorded by a thermocouple and is adjusted automatically. The furnace atmosphere may be controlled or a vacuum (10^{-3} mmHg) may be used. The spherical furnace is supplied in three modifications, a standard model for temperatures between 20 and 1000 °C, a high-temperature type for temperatures 500 − 1400 °C, and a

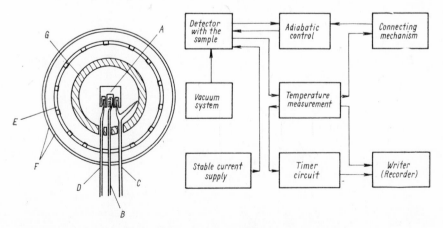

Fig. 5.8 Schematic diagram of the SHM apparatus by Rigaku Denki. *A* — sample; *B* — internal heating; *C* — thermocouple for adiabatic control; *D* — thermocouple for temperature measurement; *E* — electric furnace; *F* — shielding; *G* — adiabatic jacket

low-temperature model for temperatures from —160 °C to +200 °C (for DTA only). For DTA the sample and the reference substance are situated in metallic or quartz vessels and the differential thermocouple is placed directly in the sample or a cavity in the crucible bottom. The spherical furnace practically eliminates shifts of the baseline, owing to its symmetrical heating. In this arrangement the apparatus works in the ranges from ±50 to ±1000 μV with maximum sensitivity 1 μV. The linear heating programme is variable in the range 0.25 – 10 °C/min. In the latest model for DTA and DSC the spherical furnace has been replaced by a classical cylindrically shaped microfurnace.

The arrangement of the apparatus as an adiabatic calorimeter (SHM method) is shown in Fig. 5.8. The temperature of the sample is automatically kept the same as the temperature of the adiabatic jacket by means of a heater located inside the sample. A constant wattage (W) is used, and the time interval, Δt, during which the sample is heated by ΔT °C is measured. The specific heat c_p is proportional to the time interval, because the energy supplied to the sample is constant. If the heating lasts Δt (sec) during which the sample at T °C is heated by ΔT °C, the specific heat of the sample at T °C may be expressed as follows:

$$(c_p)_T = \frac{0.239W . \Delta t}{M . \Delta T} \tag{5.1}$$

where M is the mass of the sample.

The total transformation (or reaction) heat is determined by the area under the time versus temperature curve. The instrument works with samples of 8 cm³ at a sensitivity of 0.5 sec.

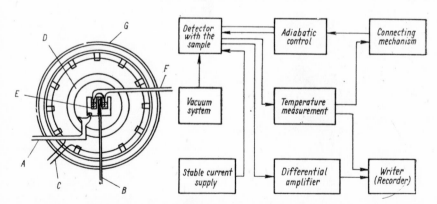

Fig. 5.9 Schematic diagram of DT analyser and calorimeter by Rigaku Denki. A — thermocouple for adiabatic control; B — internal heater; C — electric furnace; D — adiabatic shield; E — block for DTA; F — differential thermocouple; G — shielding.

The arrangement used as a differential calorimeter is shown in Fig. 5.9, This arrangement is used for quantitative studies of small energy changes and is based on the preceding two methods. Both samples are heated by independent internal heating coils. The temperature difference between the reference sample and the adiabatic jacket is followed by a differential

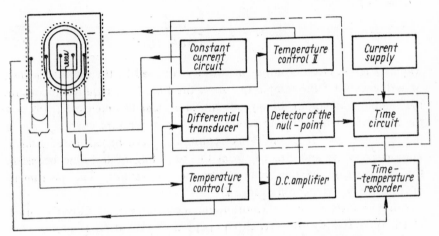

Fig. 5.10 Schematic diagram of the apparatus Shimadzu SH-2B.

thermocouple, and the furnace temperature is regulated so that this difference is eliminated. The temperature difference between the sample and the reference substance is followed by another differential thermocouple, and the output of the internal heating spiral of the reference sample is regulated in such a manner that the temperature difference is zero. Changes in the input of the internal heating spiral are recorded, and from the input of both heating spirals the energy change is estimated. With this arrangement the apparatus works in the temperature interval from -160 °C to 200 °C, or $20-500$ °C with a sensitivity of $20-300$ mW, for three ranges from 0.5 to 2.5 W for the full-scale deflection.

The Japanese firm Shimadzu Seisakusho produces, in combination with its DTA apparatus type DT 2B (p. 247), an instrument for the measurement of specific heat, type SH-2 B (Fig. 5.10). The instrument uses the principle of Sykes and Jones [30, 31], in which the sample is heated at a constant rate by an internal heater of known constant electrical input. The adiabatic block is provided with an external heating spiral. The temperature difference between the sample and the furnace is followed by a thermo-couple and is kept at zero. This arrangement is placed in another cylindrical furnace and the temperature difference betwen it and the internal adiabatic block is also kept at zero. Owing to a double adiabatic jacket, the energy supplied by the internal heater heats the sample only. An accurate timer measures and records the time necessary for the increase of the sample temperature by an amount corresponding to unit e.m.f. The values of endothermic or exothermic effects can be obtained on integration of the recorded curve. The specific heat is given by the relationship

$$c_p \cdot M + \alpha = \frac{Q \Delta t}{\Delta E (d\Theta/dE)} \qquad (5.2)$$

where Q is the electric output in calories supplied to the internal heater, M is the mass of the sample (g), Δt is the time (sec) required to elevate the sample temperature by an amount equivalent to unit e.m.f., ΔE, α is the apparatus constant, and $(d\Theta/dE)$ is the temperature characteristic of the thermocouple. If Q and M are constant, c_p can be calculated from a knowledge of Δt and ΔE, i.e. from the recording of the time necessary for the increase of unit temperature of the sample. The instrument operates in the temperature range $20-850$ °C.

Another instrument for calorimetric measurements is Calvet's calorimeter produced by the French company SETARAM. It is produced in several modifications, covering the temperature range -185 to $+800$ °C. A high-temperature version is also being developed, for temperatures up to 1300 °C. The principle of Calvet's method was given in section 3.9.7. It is used in the study of heats of mixing and of reaction heats produced between liquids or liquids and solids, as well as for the determination of combustion heats in a calorimetric bomb and for specific heats. The apparatus (type CRMT) is basically a metallic cylindrical measuring cell containing the sample. It is manufactured in two versions, the standard one of 17 mm diameter and 80 mm height, and the other larger, of 35 mm diameter, 100 mm height. The measuring cell is located symmetrically in the unit for the measurement of the heat flow, which consists of a set of thermocouples connected in series. The junctions are located alternately on the wall of the cell with the analysed sample and on the wall of the isothermal metallic jacket, where they are electrically, but not thermally, insulated. The unit is located in a metallic block, carefully thermostatically controlled. This calorimetric unit has an automatic rotating housing which mixes the substances being studied. The temperature regulator which is built into the panel unit together with the mechanical part ensures thermal stability of the jacket during the experiment. The principle of the apparatus is that the absorption or the liberation of heat in the measuring vessel causes a very slight change in its temperature, which in turn causes a heat flow between the cell and the block. The signal from the heat-flow measuring device is recorded, and by integration of this the total amount of heat may be obtained. The apparatus is calibrated either with standards (e.g. dissolution of KCl, combustion of benzoic acid) or electrically.

The preferred version of this apparatus is Calvet's microcalorimeter, based on the same principle, in which the heat flow meter also transfers heat from the containers to the block and vice versa, and induces an e.m.f. proportional to their temperature difference. This e.m.f. is at each moment proportional to the heat exchange. Thus, by using the Peltier effect, it is possible to carry out measurements when the sample itself is not a source of heat. In this apparatus two or four basic measuring units are employed which are located symmetrically within the thermostated jacket. With differential interconnection of identical cells it is possible to prevent possible temperature changes and improve the stability of the baseline. The e.m.f. produced in the thermopiles has values ranging from 10^{-8} V to several mV, and is measured by galvanometric recorders, or recorders combined with an amplifier.

In a book of this size it is impossible to describe in detail all types of commercially manufactured apparatus, because there are too many of them, they are constantly being improved, and new models are appearing all the time. Table 5.1 reviews briefly some of the other apparatus available. In this chapter only those instruments which the author considers the most interesting and useful from the technical point of view have been discussed. Special attention has been devoted to instruments aimed at miniaturization of thermal analysis, as developments in this direction are possibly the most significant.

Table 5.1

Commercially Produced Apparatus

1	2	3	4
No.	Type	Producer	Type of balance
1.	Chevenard thermobalance	A.D.A.M.E.L., 4 Passage Louis Philipe Paris XI, France	Deflection balance, photocell indication and recording on a cylindrical drum.
2.	Thermobalance Testut Serie	Testut Co., 9 Rue Brown Sequard, Paris XV, France	Compensation balance, compensation chain controlled by a servomotor, two magnetically coupled beams
3.	Thermo-Grav Aminco	Amer. Instr. Co. 8030, Georgia Ave, Silver Spring, Md., USA	Deflection balance, coil. Differential transformer
4.	Cahn electrobalance RM	Cahn Instrum. Co. 145111 Paramounth Blud., Paramounth USA	Compensation balance electromagnetic with photocell detection system
5.	Thermobalance E-H, Research Lab.	E-H Research Lab. 2161 Shattuch Ave., Berkeley, Calif. USA	Electromagnetic compensation balance
6.	Thermobalance Chyo Automatic	Chyo Balance Co., 376-2 Tsukiyama-cho Kuze, Minami-ku, Kyoto, Japan	Compensation balance (a deflection type balance is also manufactured)
7.	Quartz spring thermobalance Chyo	Chyo Balance Co., 376-2 Tsukiyama-cho Kuze, Minami-ku, Kyoto, Japan	Deflection balance
8.	Thermobalance Harrop	Harrop Precision Furnace Co., 3470 E. 5th Ave. Columbus Ohio, USA	Compensation balance
9.	Thermobalance Stone	Columbia Scientific Industries O. O. Box 6190, 3625 Bluestein Blvd., Austin. Texas, USA.	Compensation balance
10.	Thermograv. System Tem-Press	Tem-press Research Inc. 1526 William Str., St. College, Penn. USA	Compensation balance (utilisation of Cahn's thermobalance)
11.	DTA Analyser Model 1702 DR	Mitamura Rikai, Kogyo Inc., 3-2 Chome. Hongo, Bunkyo-ku, Tokyo, Japan	Automatic DTA
12	DTA Analyser	Hartmann Braun A. G. Frankfurt/Main German Fed. Republic	Photoelectrical compensation system

for Thermal Analysis

5 Location of the furnace	6 Technical parameters	7 Temperature measurement	8 Remarks
Vertical, above the balance	Capacity 10–20 g; 0.5–5 mm of record per mg	Near the crucible range 20–1200 °C	For vacuum
Vertical, above or below the balance	Capacity 200 g, range 400 mg, direct reording	Near the crucible	For vacuum and controlled atmosphere
Vertical	Capacity 20 g, range 200 mg, direct recording	Near the crucible. Double furnace, range 20–1000 °C	For vacuum and controlled atmosphere
Vertical, below the balance	Capacity 2500 mg; ranges 20 and 100 mg	Near the crucible	For vacuum and controlled atmosphere
	Capacity 200 g, range 300–3000 mg. Direct recording		
Vertical, below the balance	Capacity 160 g, sensitivity 0.1 mg, direct recording	Near the crucible, range 20–1500 °C	Work in vacuum. Simultaneous DTG and DTA recording
Vertical	Capacity 1–5 g, direct recording, error 0.5–1.0 mg	Near the crucible	For vacuum and corrosive atmospheres
Vertical, above the balance, or horizontal	Capacity 10–20 g, sensitivity 10 and 20 μg	In the sample, ranges 20–2200 °C and —40 to 500 °C	For vacuum and controlled atmosphere
Vertical, above the balance, direct recording	Capacity 1.5 g, sensitivity 2 μg. Direct recoding	In the sample, range 20–1000 °C	For vacuum and controlled atmosphere
Vertical, below the balance, direct recording	Capacity 2500 mg. Ranges 20 and 100 mg. Direct recording	Above the sample, range 20–1000 °C	For vacuum and controlled atmosphere
Horizontal furnace with a special block and measuring head, direct recording	Ranges ± 25 to ± 1000 μV, direct recording	Range 20–1250 °C	Nickel block, crucibles Pt, SiO_2, Pyrex. Also vacuum
Vertical, with various types of Ni or ceramic blocks	Broad range of sensitivity	Furnaces according to the designs of other producers	Various types, depending on the area of use

Table 5.1

1	2	3	4
No.	Type	Producer	Type of the apparatus
13.	DTA device Stone	Robert L. Stone Co. Automatic DTA 3314 Westhill Drive, Austin 4, Texas, USA	Automatic DTA
14.	DTA Analyser Deltatherm	Technical Equipm. Corp., 917 Acoma Str., Denver 4, Colorado, USA	Automatic DTA
15.	High-temperature DTA Analyser	Rempipnal, Volgograd	Automatic DTA
16.	DTA Analyser "D-T-A-1"	VEB Elektro, Bad Franken-hausen	Automatic DTA
17.	Micro DTA Ana-lyser, Model M3	Bureau de liaison, 113 Rue de l'Université, Paris 7	Automatic DTA
18.	Automatic deri-vation and diffe-rential thermo-analyser Series T-1 Series TA-1 Series TR-160	VOIAND Corporation 27 Centre Ave., New Rockell, N.Y., USA	Compensation balance with simultaneous DTA System: TGA + DTGA + + DTA

(continued)

5 Location of the furnace	6 Technical parameters	7 Temperature measurement	8 Remarks
Vertical block from various materials and with cavities for 2–9 samples		Various furnaces for temperatures up to 1400 °C	
		Various furnaces for temperatures up to 1600 °C	Vacuum or dynamic atmosphere and increased pressure
With cavities for 2–9 samples			
Vertical arrangement of the block	Temperature range 20–2800 °C	By optical pyrometer	Produced according to ref. [32]
Vertical arrangement, various types of blocks	Temperature range 20–1200 °C Direct current amplifier		Two exchangeable furnaces
Vertical arrangement	Three types of exchangeable probes for various quantities of the sample. Temperature range 20–1500 °C		Macro and micro method
Horizontal arrangement of sample holder	Temperature range 20–1600 °C. Balance capacity 1 g, sensitivity 0.1 mg. Ranges of DTA: 10–1000 μV	Various types of furnaces	A series of types is produced: T-1R, T-1A, T-1RU, T-1RA, T-1RUA; also with a spring balance in Series SL

References

1. BLAŽEK A., HALOUSEK J. *Silikáty* **6**, 100 (1962).
2. BLAŽEK A. *Silikáty* **4**, 52 (1960).
3. BLAŽEK A. *Bergakademie* **12**, 191 (1960).
4. BLAŽEK A. *Silikáty* **1**, 158 (1957).
5. BLAŽEK A. *Hutnické Listy* **12**, 1096 (1957).
6. BLAŽEK A. *Hutnické Listy* **13**, 505 (1958).
7. BLAŽEK A., HALOUSEK J. *Hutnické Listy* **14**, 244 (1959).
8. BLAŽEK A., HALOUSEK J. *Automatic Balance,* Czechoslov. Pat. No. 92079 (1957).
9. BERÁNEK M. *Silikáty* **10**, 93 (1966).
10. BERÁNEK M. *Report of the Research Institute of Low-Tension Technique Tesla,* Prague, No. 13045/3 (1963); No. 13059/3 (1963).
11. BOERSMA S. L. *J. Am. Ceram. Soc.* **38**, 281 (1955).
12. DUVAL C. *Inorganic Thermogravimetric Analysis,* 2nd Ed. Elsevier, London 1964.
13. GARN P. D. *Anal. Chem.* **33**, 1247 (1961).
14. GORDON S., CAMPBELL C. *Anal. Chem.* **32**, 271 R (1960).
15. JACQUE L., GUIOCHON G., GENDREL P. *Bull. Soc. Chim. France* 1061 (1961).
16. KULP J. L., KERR P. E. *Science* **105**, 413 (1947).
17. LEWIN S. Z. *J. Chem. Educ.* **39**, A 575 (1962).
18. MAZIÈRES C. *Compt. Rend.* **248**, 2990 (1959).
19. MURPHY C. B. *Anal. Chem.* **34**, 298 R (1962).
20. MURPHY C. B. *Anal. Chem.* **32**, 168 R (1960).
21. MURPHY C. B. *Anal. Chem.* **38**, 443 R (1966).
22. NAGASAKI S., TAKIGI T. *Appl. Physics (Japan)* **1B**, 105 (1948).
23. PAKULAK J. M., LEONARD G. W. *Anal. Chem.* **31**, 1037 (1959).
24. PAULIK F., PAULIK J., ERDEY L. *Derivatograph. Information Booklet of Metrimpex Company, Budapest* 1965.
25. PAULIK F., PAULIK J., ERDEY L. *Z. Anal. Chem.* **160**, 241 (1958).
26. PAULIK F., PAULIK J., GÁL S. *Z. Anal. Chem.* **163**, 321 (1961).
27. PAULIK J. *Proc. Anal. Chem. Conference, Budapest* 1966, Vol. 3, 339.
28. PAULIK F. *Proc. Anal. Chem. Conference, Budapest* 1966, Vol. 3, 333.
29. PAULIK F., PAULIK J., ERDEY L. *Z. Anal. Chem.* **160**, 321 (1958).
30. SYKES C. *Proc. Roy. Soc.* **148**, 422 (1935).
31. SYKES C., JONES F. W. *J. Inst. Metals* **59**, 257 (1936).
32. NEDUMOV N. A. *Proc. First Intern. Conf. Thermal Analysis, Aberdeen.* Macmillan, London 1965, Paper 1.4., p. 8.
33. Neue thermogravimetrische Analysengeräte. Thermomat S. *J. Thermal Anal.* **2**, 209 (1970).

APPENDIX I

TABLES OF THERMAL E.M.F. VALUES OF SOME IMPORTANT THERMOCOUPLES [1, 2]

Table 1

Thermocouple Pt 10%Rh–Pt Type: Le Chatelier

Thermal e.m.f. in mV
Temperature of reference sites: 0 °C

°C	0	1	2	3	4	5	6	7	8	9
0	0.000	0.005	0.011	0.016	0.022	0.028	0.033	0.039	0.044	0.050
10	.056	.0,61	.0.73	.0,78	.078	.084	.090	.096	.102	.107
20	.113	.119	.125	.131	.137	.143	.149	.155	.161	.167
30	.173	.179	.185	.191	.198	.204	.210	.216	.222	.229
40	.235	.241	.247	.254	.260	.266	.273	.279	.286	.292
50	.299	.305	.312	.318	.325	.331	.338	.344	.351	.357
60	.364	.371	.377	.384	.391	.397	.404	.411	.418	.425
70	.431	.438	.445	.452	.459	.466	.473	.479	.486	.493
80	.500	.507	.514	.521	.528	.535	.543	.550	.557	.564
90	.571	.578	.585	.593	.600	.607	.614	.621	.629	.636
100	.643	.651	.658	.665	.673	.680	.687	.694	.702	.709
110	.717	.724	.732	.739	.747	.754	.762	.769	.777	.784
120	.792	.800	.807	.815	.823	.830	.838	.845	.853	.861
130	.869	.876	.884	.892	.900	.907	.915	.923	.931	.939
140	.946	.962	.954	.970	.978	.986	.994	1.002	1.009	1.017
150	1.025	1.033	1.041	1.049	1.057	1.065	1.073	1.081	1.089	1.097
160	1.106	1.114	1.122	1.130	1.138	1.146	1.154	1.162	1.170	1.179
170	1.187	1.915	1.203	1.211	1.220	1.228	1.236	1.244	1.253	1.261
180	1.269	1.277	1.286	1.294	1.302	1.311	1.319	1.327	1.336	1.344
190	1.352	1.361	1.369	1.377	1.386	1.394	1.403	1.411	1.419	1.428
200	1.436	1.445	1.453	1.462	1.470	1.479	1.487	1.496	1.504	1.513
210	1.521	1.530	1.538	1.547	1.555	1,564	1.573	1.581	1.590	1.598
220	1.607	1.615	1.624	1.633	1.641	1.650	1.659	1.667	1.676	1.685
230	1.693	1.702	1.710	1.719	1.728	1.736	1.745	1.754	1.763	1.771
240	1.780	1.789	1.798	1.805	1.815	1.824	1.833	1.841	1.850	1.859
250	1.868	1.877	1.885	1.894	1.903	1.912	1.921	1.930	1.938	1.947
260	1.956	1.965	1.974	1.983	1.992	2.001	2.009	2.018	2.027	2.036
270	2.045	2.054	2.063	2.072	2.081	2.090	2.099	2.108	2.117	2.126
280	2.135	2.144	2.153	2.162	2.171	2.180	2.189	2.198	2.207	2.216
290	2.225	2.234	2.243	2.252	2.261	2.271	2.280	2.289	2.298	2.307

Table 1

°C	0	1	2	3	4	5	6	7	8	9
300	2.316	2.235	2.334	2.343	2.353	2.362	2.371	2.380	2.389	2.398
310	2.408	2.417	2.426	2.435	2.444	2.453	2.463	2.472	2.481	2.490
320	2.499	2.509	2.518	2.527	2.536	2.546	2.555	2.564	2.573	2.583
330	2.592	2.601	2.610	2.620	2.629	2.638	2.648	2.657	2.666	2.676
340	2.685	2.694	2.704	2.713	2.722	2.731	2.741	2.750	2.760	2.769
350	2.778	2.788	2.797	2.806	2.816	2.825	2.834	2.844	2.853	2.863
360	2.872	2.881	2.891	2.900	2.910	2.919	2.929	2.938	2.947	2.957
370	2.966	2.976	2.985	2.995	3.004	3.014	3.023	3.032	3.042	3.051
380	3.061	3.070	3.080	3.089	3.099	3.108	3.118	3.127	3.137	3.146
390	3.156	3.165	3.175	3.184	3.194	3.203	3.213	3.222	3.232	3.241
400	3.251	3.261	3.270	3.280	3.289	3.299	3.308	3.318	3.327	3.337
410	3.347	3.356	3.366	3.375	3.385	3.394	3.404	3.414	3.423	3.433
420	3.442	3.452	3,462	3.471	3.481	3.490	3.500	3.510	3.519	3.529
430	3.539	3.548	3.558	3.567	3.577	3.587	3.596	3.606	3.616	3.625
440	3.635	3.645	3.654	3.664	3.674	3.683	3.693	3.703	3.712	3.722
450	3.732	3.741	3.751	3.761	3.771	3.780	3.790	3.800	3.809	3.819
460	3.829	3.839	3.848	3.858	3.868	3.878	3.887	3.897	3.907	3.917
470	3.926	3.936	3.946	3.956	3.965	3.975	3.985	3.995	4.004	4.014
480	4.024	4.034	4.044	4.053	4.063	4.073	4.083	4.093	4.103	4.112
490	4.122	4.132	4.142	4.142	4.162	4.171	4.181	4.191	4.201	4.211
500	4.221	4.230	4.240	4.250	4.260	4.270	4.280	4.290	4.300	3.310
510	4.319	4.329	4.339	4.349	4.359	4.369	4.379	4.389	4.399	4.409
520	4.419	4.428	4.438	4.448	4.458	4.468	4.478	4.488	4.498	4.508
530	4.518	4.528	4.538	4.548	4.558	4.568	4.578	4.588	4.598	4.608
540	4.618	4.628	4.638	4.648	4.658	4.668	4.678	4.688	4.698	4.708
550	4.718	4.728	4.738	4.748	4.758	4.768	4.778	4.788	4.798	4.808
560	4.818	4.828	4.839	4.849	4.859	4.869	4.879	4.889	4.899	4.909
560	4.919	4.929	4.939	4.950	4.960	4.970	4.980	4.990	5.000	5.010
580	5.020	5.031	5.041	5.051	5.061	5.071	5.081	5.091	5.102	5.112
590	5.112	5.132	5.142	5.151	5.163	5.173	5.183	5.193	5.203	5.214
600	5.224	5.234	5.244	5.254	5.265	5.275	5.285	5.295	5.306	5.316
610	5.326	5.336	5.346	5.357	5.367	5.377	5.388	5.398	5.408	5.418
620	5.429	5.439	5.449	5.459	5.470	5.480	5.490	5.501	5.511	5.521
630	5.532	5.542	5.552	5.563	5.573	5.583	5.593	5.604	5.614	5.624
640	5.635	5.645	5.655	5.666	5.676	5.686	5.707	5.707	5.717	5.728
650	5.738	5.748	5.759	5.769	5.779	5.790	5.800	5.811	5.821	5.831
660	5.842	5.852	5.862	5.873	5.883	5.894	5.904	5.914	5.925	5.935
670	5.946	5.956	5.967	5.977	5.987	5.998	6.008	6.019	6.029	4.040
680	6.050	6.060	6.071	6.081	6.092	6.102	6.113	6.123	6.134	6.144
690	6.155	6.165	6.176	6.186	6.197	6.207	6.218	6.228	6.239	6.249
700	6.260	6.270	6.281	6.291	6.302	6.312	6.323	6.333	6.344	6.355
710	6.365	6.376	6.386	6.397	6.407	6.418	6.429	6.439	6.450	6.460
720	6.471	6.481	6.492	6.503	6.513	6.524	6.534	6.545	6.556	6.566
730	6.577	6.588	6.598	6.609	6.619	6.630	6.641	6.651	6.662	6.673
740	6.683	6.694	6.705	6.715	6.726	6.737	6.747	6.758	6.769	6.779

(continued)

°C	0	1	2	3	4	5	6	7	8	9
750	6.790	6.801	6.811	6.822	6.833	6.844	6.854	6.865	6.876	6.886
760	6.897	6.908	6.919	6.929	6.940	6.951	6.962	6.972	6.983	6.994
770	7.005	7.015	7.026	7.037	7.047	7.058	7.069	7.080	7.091	7.102
780	7.112	7.123	7.134	7.145	7.156	7.166	7.177	7.188	7.199	7.210
790	7.220	7.231	7.242	7.253	7.264	7.275	7.286	7.296	7.307	7.318
800	7.329	7.340	7.351	7.362	7.372	7.383	7.394	7.405	7.416	7.427
810	7.438	7.449	7.460	7.470	7.481	7.492	7.503	7.514	7.525	7.536
820	7.547	7.558	7.569	7.580	7.591	7.602	7.613	7.623	7.634	7.645
830	7.656	7.667	7.678	7.689	7.700	7.711	7.722	7.733	7.744	7.755
840	7.766	7.777	7.788	7.799	7.810	7,821	7.832	7.843	7.854	7.865
850	7.876	7.887	7.898	7.910	7.921	7.932	7.943	7.954	7.965	7.976
860	7.987	7.998	8.009	8.020	8.031	8.042	8.053	8.064	8.076	8.087
870	8.098	8.109	8.120	8.131	8.142	8.153	8.164	8.176	8.187	8.198
880	8.209	8.220	8.231	8.242	8.254	8.265	8.276	8.287	8.298	8.309
890	8.320	8.332	8.343	8.354	8.365	8.376	8.388	8.399	8.410	8.421
900	8.432	8.444	8.455	8.466	8.477	8.488	8.500	8.511	8.522	8.533
910	8.545	8.556	8.567	8.578	8.590	8.601	8.612	8.623	8.635	8.646
920	8.657	8.668	8.680	8.691	8.702	8.714	8.725	8.736	8.747	8.759
930	8.770	8.781	8.793	8.804	8.815	8.827	8.838	8.849	8.861	8.872
940	8.883	8.895	8.906	8.917	8.929	8.940	8.951	8.963	8.974	8.986
950	8.997	9.008	9.020	9.031	9.042	9.054	9.065	9.077	9.088	9.099
960	9.111	9.122	9.134	9.145	9.157	9.168	9.179	9.191	9.202	9.214
970	9.225	9.236	9.248	9.260	9.271	9.282	8.294	9.305	9.317	9.328
980	9.340	9.351	9.363	9.374	9.386	9.397	9.409	9.420	9.432	9.443
990	9.455	9.466	9.478	9.489	9.501	9.512	9.524	9.535	9.547	9.559
1000	9.570	9.582	9.393	9.605	9.616	9.628	9.639	9.651	9.663	9.674
1010	9.686	9.697	9.709	9.720	9.732	9.744	9.755	9.767	9.779	9.790
1020	9.802	9.813	9.825	9.837	9.848	9.860	9.871	9.883	9.895	9.906
1030	9.918	9.930	9.941	9.953	9.965	9.976	9.988	10.000	10.011	10.023
1040	10.035	10.046	10.058	10.070	10.082	10.093	10.105	10.117	10.128	10.140
1050	10.152	10.163	10.175	10.187	10.199	10.210	10.222	10.234	10.246	10.257
1060	10.269	10.281	10.293	10.304	10.316	10.328	10.340	10.351	10.363	10.375
1070	10.387	10.399	10.410	10.422	10.434	10.446	10.458	10.469	10.481	10.493
1080	10.505	10.517	10.528	10.540	10.552	10.564	10.576	10.587	10.599	10.611
1090	10.623	10.635	10.647	10.658	10.670	10.682	10.694	10.706	10.718	10.729
1100	10.741	10.753	10.765	10.777	10.789	10.801	10.812	10.824	10.836	10.848
1110	10.860	10.872	10.884	10.896	10.907	10.919	10.931	10.943	10.955	10.967
1120	10.979	10.991	11.003	11.014	11.026	11.038	11.050	110.62	11.074	11.086
1130	11.098	11.110	11.122	11.133	11.145	11.157	11.169	11.181	11.193	11.205
1140	11.217	11.229	11.241	11.253	11.265	11.277	11.289	11.300	11.312	11.324
1150	11.336	11.348	11.360	11.372	11.384	11.396	11.408	11.420	11.342	11.444
1160	11.456	11.468	11.480	11.492	11.504	11.516	11.528	11.540	11.552	11.564
1170	11.575	11.587	11.599	11.611	11.623	11.623	11.647	11.659	11.671	11.683
1180	11.695	11.707	11.719	11.731	11.743	11.755	11.767	11.779	11.791	11.803
1190	11.815	11.827	11.839	11.851	11.863	11.875	11.887	11.899	11.911	11.923

Table 1

°C	0	1	2	3	4	5	6	7	8	9
1200	11.935	11.947	11.959	11.971	11.983	11.995	12.007	12.019	12.031	12.043
1210	12.055	12.067	12.079	12.091	12.103	12.115	12.127	12.139	12.151	12.163
1220	12.175	12.187	12.200	12.212	12.224	12.236	12.248	12.260	12.272	12.284
1230	12.296	12.308	12.320	12.322	12.344	12.356	12.368	12.380	12.392	12.404
1240	12.416	12.428	12.440	12.452	12.464	12.476	12.488	12.500	12.512	12.524
1250	12.536	12.548	12.560	12.573	12.585	12.597	12.609	12.621	12.633	12.645
1260	12.657	12.669	12.681	12.693	12.705	12.717	12.729	12.741	12.753	12.765
1270	12.777	12.789	12.801	12.813	12.825	12.837	12.849	12.861	12.873	12.885
1280	12.897	12.909	12.921	12.933	12.945	12.957	12.969	12.981	12.993	13.005
1290	13.018	13.030	13.042	13.054	13.066	13.078	13.090	13.102	13.141	13.126
1300	13.138	13.150	14.162	13.174	13.186	13.198	13.210	13.222	13.234	13.246
1310	13.258	13.270	13.282	13.294	13.306	13.318	13.330	13.342	13.354	13.366
1320	13.378	13.390	13.402	13.414	13.426	13.438	13.450	13.462	13.474	13.486
1330	13.498	13.510	13.522	13.534	13.546	13.558	13.570	13.582	13.594	13.606
1340	13.618	13.630	13.642	13.654	13.666	13.678	13.690	13.702	13.714	13.726
1350	13.738	13.750	13.762	13.774	13.786	13.798	13.810	13.822	13.834	13.846
1360	13.858	13.870	13.882	13.894	13.906	13.918	13.930	13.942	13.954	13.966
1370	13.978	13.990	14.002	14.014	14.026	14.038	14.050	14.062	14.074	14.086
1380	14.098	14.110	14.122	14.133	14.145	14.157	14.169	14.181	14.193	14.205
1390	14.217	14.229	14.241	14.253	14.265	14.277	14.289	14.301	14.313	14.325
1400	14.337	14.349	14.361	14.373	14.385	14.397	14.409	14.421	14.433	14.455
1410	14.457	14.469	14.481	14.493	14.504	14.516	14.528	14.540	14.552	14.564
1420	14.576	14.588	14.600	14.612	14.624	14.636	14.648	14.660	14.672	14.684
1430	14.696	14.708	14.720	14.732	14.744	14.755	14.767	14.779	14.791	14.803
1440	14.815	14.827	14.839	14.851	14.863	14.875	14.887	14.899	14.911	14.923
1450	14.935	14.946	14.958	14.970	14.982	14.994	15.006	15.018	15.030	15.042
1460	15.054	15.066	15.078	15.090	15.102	15.113	15.125	15.137	15.149	15.161
1470	15.173	15.185	15.197	15.209	15.221	15.233	15.245	15.256	15.268	15.280
1480	15.292	15.304	15.316	15.328	15.340	15.352	15.364	15.376	15.387	15.399
1490	15.411	15.423	15.435	15.447	15.459	15.471	15.483	15.495	15.507	15.518
1500	15.530	15.542	15.554	15.566	15.578	15.590	15.609	15.614	15.625	15.637
1510	15.649	15.661	15.673	15.685	15.697	15.709	15.721	15.732	15.744	15.756
1520	15.768	15.780	15.792	15.804	15.816	15.827	15.839	15.851	15.863	15.875
1530	15.887	15.899	15.911	15.922	15.934	15.946	15.958	15.970	15.982	15.994
1540	16.006	16.017	16.029	16.041	16.053	16.065	16.077	16.089	16.100	16.112
1550	16.124	16.136	16.148	16.160	16.171	16.183	16.195	16.207	16.219	16.231
1560	16.243	16.254	16.266	16.278	16.290	16.302	16.314	16.325	16.337	16.349
1570	16.361	16.373	16.385	16.396	16.408	16.420	14.632	16.444	16.456	16.467
1580	16.479	16.491	16.503	16.515	16.527	16.538	16.550	16.562	16.574	16.586
1590	16.597	16.609	16.621	16.633	16.645	16.657	16.668	16.680	16.692	16.704
1600	16.716	16.727	16.739	16.751	16.763	16.775	16.786	16.798	16.810	16.822
1610	16.834	16.845	16.857	15.869	16.881	16.893	16.904	16.916	16.928	16.940
1620	16.952	16.963	16.975	16.987	16.999	17.010	17.022	17.034	17.046	17.058
1630	17.069	17.081	17.093	17.105	17.116	17.128	17.140	17.152	17.163	17.175
1640	17.187	17.199	17.211	17.222	17.234	17.246	17.258	17.269	17.281	17.293

(continued)

°C	0	1	2	3	4	5	6	7	8	9
1650	17.305	17.316	17.328	17.340	17.352	17.363	17.375	17.387	17.398	17.410
1660	17.422	17.434	17.446	17.457	17.469	17.481	17.492	17.504	17.516	17.528
1670	17.539	17.551	17.563	17.575	17.586	17.598	17.610	17.621	17.633	17.645
1680	17.657	17.668	17.680	17.692	17.704	17.715	17.727	17.739	17.750	17.762
1690	17.774	17.785	17.797	17.809	17.821	17.832	17.844	17.856	17.867	17.879
1700	17.891	17.902	17.914	17.926	17.938	17.949	17.961	17.973	17.984	17.996
1710	18.008	18.019	18.031	18.043	18.054	18.066	18.078	18.089	18.101	18.113
1720	18.124	18.136	18.148	18.159	18.171	18.183	18.194	18.206	18.218	18.229
1730	18.241	18.253	18.264	18.276	18.288	18.299	18.311	18.323	18.334	18.346
1740	18.358	18.369	18.381	18.393	18.404	18.416	18.427	18.439	18.451	18.462
1750	18.474	18.486	18.497	18.509	18.520	18.532	18.544	18.555	18.567	18.579
1760	18.590	18.602	18.613	18.625	18.637	18.648	18.660	18.672	18.683	18.695

Correction for the cold end of the Pt—Pt 10%Rh thermocouple

°C	0	10	20 Millivolts	30	40
0	0.000	0.055	0.112	0.172	0.233
1	0.005	0.060	0.118	0.178	0.239
2	0.011	0.066	0.124	0.184	0.246
3	0.016	0.072	0.130	0.190	0.252
4	0.022	0.078	0.136	0.196	0.258
5	0.027	0.083	0.142	0.202	0.265
6	0.033	0.089	0.148	0.208	
7	0.038	0.095	0.154	0.214	
8	0.044	0.100	0.160	0.221	
9	0.049	0.106	0.166	0.227	

Table 2

Thermocouple Pt 6% Rh—Pt 30%   Type: PtRh "El 18"

Thermal e.m.f. in mV
Temperature of reference sites: 0 °C

°C	mV	mV/°C	°C	mV	mV/°C	°C	mV	mV/°C
0	0		490	1.202	0.0050	980	4.677	0.0097
10	—0.002	0.0001	500	1.252	0.0051	990	4.767	0.0092
20	—0.003	0.0001	510	1.303	0.0052	1000	4.859	0.0092
30	—0.002	0.0001	520	1.355	0.0053	1010	4.951	0.0092
40	—0.001	0.0003	530	1.408	0.0054	1020	4.951	0.0092
50	—0.002	0.0004	540	1.462	0.0055	1030	5.136	0.0094
60	0.006	0.0005	550	1.517	0.0055	1040	5.230	0.0095
70	0.011	0.0007	560	1.572	0.0056	1050	5.325	0.0095
80	0.018	0.0007	570	1.628	0.0058	1060	5.420	0.0096
90	0.025	0.0010	580	1.686	0.0058	1070	5.516	0.0097
100	0.035	0.0009	590	1.744	0.0059	1080	5.613	0.0097
110	0.044	0.0011	600	1.803	0.0060	1090	5.710	0.0098
120	0.055	0.0012	610	1.863	0.0061	1100	5.808	0.0098
130	0.067	0.0013	620	1.924	0.0062	1110	5.906	0.0099
140	0,080	0.0014	630	1.986	0.0063	1120	6,005	0.0100
150	0,094	0.0015	640	2.049	0.0064	1130	6.105	0.0100
160	0,109	0.0017	650	2.113	0.0065	1140	6.205	0.0101
170	0.126	0.0018	660	2.178	0.0065	1150	6.306	0.0102
180	0.144	0.0018	670	2.243	0.0066	1160	6.408	0.0102
190	0.162	0.0019	680	2.309	0.0067	1170	6.510	0.0103
200	0.181	0.0021	690	2.376	0.0068	1180	6.613	0.0103
210	0.202	0.0022	700	2.444	0.0069	1190	6.716	0.0103
220	0.224	0.0023	710	2.513	0.0069	1200	6.819	0.0104
230	0.247	0.0024	720	2.582	0.0071	1210	6.923	0.0104
240	0,271	0.0025	730	2.653	0.0072	1220	7.027	0.0106
250	0.296	0.0026	740	2.725	0.0072	1230	7.133	0.0106
260	0.322	0.0027	750	2.797	0.0073	1240	7.239	0.0106
270	0.349	0.0028	760	2.870	0.0074	1250	7.345	0.0106
280	0.377	0.0029	770	2.944	0.0075	1260	7.451	0.0107
290	0.406	0.0030	780	3.019	0.0075	1270	7.558	0.0107
300	0.436	0.0031	790	3.094	0.0077	1280	7.665	0.0108
310	0.467	0.0033	800	3.171	0.0077	1290	7.773	0.0109
320	0.500	0.0033	810	3.248	0.0078	1300	7.882	0.0109
330	0.533	0.0034	820	3.326	0.0079	1310	7.991	0.0109
340	0.567	0.0036	830	3.405	0.0079	1320	8.100	0.0109
350	0.603	0.0036	840	3.484	0,0080	1330	8.209	0.0110
360	0.639	0.0037	850	3.564	0,0081	1340	8.319	0.0110
370	0.676	0.0039	860	3.645	0,0082	1350	8.429	0.0110
380	0.715	0.0039	870	3.727	0.0083	1360	8.539	0.0111
390	0.754	0.0041	880	3.810	0.0083	1370	8.650	0.0111
400	0.795	0.0041	890	3.893	0.0085	1380	8.761	0.0112
410	0.836	0.0042	900	3.987	0.0085	1390	8.873	0.0112
420	0.878	0.0043	910	4.063	0.0085	1400	8.985	0.0112
430	0.921	0.0045	920	4.148	0.0056	1410	9.097	0.0112
440	0.966	0.0045	930	4.234	0.0087	1420	9.209	0.0112
450	1.011	0.0046	940	4.321	0.0088	1430	9.321	0.0112
460	1.057	0.0048	950	4.409	0.0089	1440	9.433	0.0112
470	1.105	0.0048	960	4.498	0.0089	1450	9.545	0.0113
480	1.153	0.0049	970	4.587	0.0090	1460	9.658	0.0113

Table 2 (continued)

°C	mV	mV/°C	°C	mV	mV/°C	°C	mV	mV/°C
1470	9.771	0.0113	1590	11.125	0.0113	1710	12.476	0.0112
1480	9.884	0.0131	1600	11.238	0.0113	1720	12.588	0.0113
1490	9.997	0.0113	1610	11.351	0.0112	1730	12.701	0.0112
1500	10.110	0.0113	1620	11.463	0.0113	1740	12.813	0.0113
1510	10.223	0.0113	1630	11.576	0.0112	1750	12.926	0.0112
1520	10.336	0.0113	1640	11.688	0.0113	1760	13.038	0.0113
1530	10.449	0.0112	1650	17.801	0.0112	1770	13.151	0.0112
1540	10.561	0.0113	1660	11.913	0.0113	1780	12.263	0.0113
1550	10.674	0.0113	1670	12.026	0.0112	1790	13.375	0.0112
1560	10.787	0.0113	1680	12.138	0.0113	1800	13.488	.
1570	10.900	0.0112	1690	12.251	0.0112			
1580	11.012	0.0113	1700	12.363	0.0113			

Table 3

Thermocouple Pt 13% Rh—Pt

Thermal e.m.f. in mV
Temperature of reference sites: 0 °C

°C	mV	mV/°C	°C	mV	mV/°C	°C	mV	mV/°C
0	0	0.0054	540	4.905	0.0112	1080	11.601	0.0135
10	0.054	0.0057	550	5.017	0.0112	1090	11.736	0.0135
20	0.111	0.0059	560	5.129	0.0113	1100	11.871	0.0136
30	0.170	0.0061	570	5.242	0.0113	1110	12.007	0.0136
40	0.231	0·0064	580	5.355	0.0114	1120	12.143	0.0136
50	0.295	0.0066	590	6.469	0.0114	1130	12.279	0.0136
60	0.361	0.0068	600	5.583	0.0114	1140	12.415	0.0136
70	0.429	0.0070	610	5.697	0.0115	1150	12.551	0.0137
80	0.499	0.0072	620	5.812	0.0116	1160	12.688	0.0137
90	0.571	0.0073	630	5.928	0.0116	1170	12.825	0.0138
100	0.644	0.0075	640	6.044	0.0117	1180	12.963	0.0138
110	0.719	0.0077	650	6.161	0.0118	1190	13.101	0.0139
120	0.796	0.0079	660	6.279	0.0118	1200	13.240	0.0139
130	0.875	0.0080	670	6.397	0.0118	1210	13.379	0.0139
140	0.955	0.0081	680	6.515	0.0119	1220	13.518	0.0139
150	1.036	0.0083	690	6.634	0.0120	1230	13.657	0.0139
160	1.119	0.0084	700	6.754	0.0120	1240	13.796	0.0139
170	1.203	0.0086	710	6.874	0.0120	1250	13.935	0.0139
180	1.289	0.0086	720	6.994	0.0121	1260	14.074	0.0140
190	1.375	0.0088	730	7.115	0.0121	1270	14.214	0.0140
200	1.463	0.0089	740	7.236	0.0122	1280	14.354	0.0140
210	1.552	0.0089	750	7.358	0.0122	1290	14.494	0.0140
220	1.641	0.0091	760	7.840	0.0122	1300	14.634	0.0141
230	1.732	0.0091	770	7.602	0.0123	1310	14.775	0.0141
240	1.823	0.0093	780	7.725	0.0123	1320	14.916	0.0141
250	1.916	0.0094	790	7.848	0.0124	1330	15.057	0.0141
260	2.010	0.0094	800	7.972	0.0124	1340	15.198	0.0141
270	2.104	0.0095	810	8.096	0.0125	1350	15.339	0.0142
280	2.199	0.0096	820	8.221	0.0125	1360	15.481	0.0142
290	2.295	0.0097	830	8.346	0.0126	1370	15.623	0.0142
300	2.392	0.0097	840	8.472	0.0126	1380	15.765	0.0142
310	2.489	0.0098	850	8.598	0.0127	1390	15.907	0.0142
320	2.587	0.0099	860	8.725	0.0127	1400	16.049	0.0142
330	2.686	0.0100	870	8.852	0.0127	1410	16.191	0.0142
340	2.786	0.0100	880	8.979	0.0127	1420	16.333	0.0142
350	2.886	0.0101	890	9.106	0.0127	1430	16.475	0.0142
360	2.987	0.0101	900	9.233	0.0128	1440	16.617	0.0142
370	3.088	0.0102	910	9.361	0.0128	1450	15.759	0.0142
380	3.190	0.0103	920	9.489	0.0129	1460	16.901	0.0142
390	3.293	0.0103	930	9.618	0.0129	1470	17.043	0.0142
400	3.396	0.0104	940	9.747	0.0130	1480	17.185	0.0142
410	3.500	0.0105	950	9.877	0.0130	1490	17.327	0.0140
420	3.605	0.0105	960	10.007	0.0131	1500	17.467	0.0140
430	3.710	0.0106	970	10.138	0.0132	1510	17.607	0.0140
440	3.816	0.0106	980	10.270	0.0132	1520	17.747	0.0139
450	3.922	0.0107	990	10.402	0·0132	1530	17.886	0.0139
460	4.029	0.0108	1000	10.534	0.0132	1540	18.025	0.0139
470	4.137	0.0108	1010	10.666	0.0133	1550	18.164	0.0139
480	4.245	0.0108	1020	10.799	0.0133	1560	18.303	0.0138
490	4.353	0.0109	1030	10.932	0.0133	1570	18.441	0.0137
500	4.462	0.0110	1040	11.065	0.0133	1580	18.578	0.0137
510	4.572	0.0111	1050	11.198	0.0134	1590	18.715	0.0173
520	4.683	0.0111	1060	11.332	0.0134	1600	18.852	
530	4.794	0.0111	1070	11.466	0.0135			

Table 4

Thermocouple Au40%Pd–Pt10%Rh   Type: Pallaplat

Thermal e.m.f. in mV
Temperature of reference sites: 0 °C

°C	mV	mV/°C	°C	mV	mV/°C	°C	mV	mV/°C
0	0	0.025	440	16.93	0.047	880	39.32	0.052
10	0.25	0.025	450	17.40	0.047	890	39.84	0.052
20	0.50	0.025	460	17.87	0.047	900	40.36	0.052
30	0.75	0.027	470	18.34	0.047	910	40.88	0.052
40	1.02	0.028	480	18.81	0.048	920	41.40	0.052
50	1.30	0.030	490	19.29	0.048	930	41.92	0.052
60	1.60	0.030	500	19.77	0.048	940	42.44	0.052
70	1.90	0.032	510	20.25	0.048	950	42.96	0.052
80	2.22	0.032	520	20.73	0.048	960	43.48	0.051
90	2.54	0.032	530	21.21	0.049	970	43.99	0.051
100	2.86	0.034	540	21.70	0.049	980	44.50	0.057
110	3.20	0.034	550	22.19	0.050	990	45.01	0.051
120	3.54	0.035	560	22.69	0.050	1000	45.52	0.051
130	3.89	0.036	570	23.19	0.051	1010	46.03	0.051
140	4.25	0.036	580	23.70	0.051	1020	46.54	0.050
150	4.61	0.036	590	24.21	0.051	1030	47.04	0.050
160	4.97	0.037	600	24.72	0.051	1040	47.54	0.050
170	5.34	0.038	610	25.23	0.051	1050	48.04	0.050
180	5.72	0.039	620	25.74	0.051	1060	48.54	0.050
190	6.11	0.039	630	26.25	0.051	1070	49.04	0.050
200	6.50	0.040	640	26.76	0.051	1080	49.54	0.050
210	6.90	0.041	650	27.27	0.051	1090	50.04	0.050
220	7.31	0.041	660	27.78	0.051	1100	50.54	0.050
230	7.72	0.041	670	28.29	0.052	1110	51.04	0.050
240	8.13	0.041	680	28.81	0.052	1120	51.54	0.050
250	8.54	0.041	690	29.33	0.052	1130	52.04	0.049
260	8.95	0.041	700	29.85	0.052	1140	52.53	0.049
270	9.36	0.041	710	30.37	0.052	1150	53.02	0.049
280	9.77	0.041	720	30.89	0.052	1160	53.51	0.049
290	10.18	0.042	730	31.41	0.052	1170	54.00	0.049
300	10.60	0.042	740	31.93	0.053	1180	54.49	0.049
310	11.02	0.043	750	32.46	0.053	1190	54.98	0.049
320	11.45	0.043	760	32.99	0.053	1200	55.47	0.050
330	11.88	0.044	770	33.52	0.053	1210	55.97	0.049
340	12.32	0.045	780	43.05	0.053	1220	56.46	0.049
350	12.77	0.045	790	34.58	0.053	1230	56.95	0.049
360	13.22	0.045	800	35.11	0.053	1240	57.44	0.049
370	13.67	0.046	810	35.64	0.053	1250	57.93	0.049
380	14.13	0.046	820	36.17	0.053	1260	58.42	0.049
390	14.59	0.046	830	36.70	0.053	1270	58.91	0.049
400	15.05	0.047	840	37.23	0.053	1280	59.40	0.049
410	15.52	0.047	850	37.76	0.052	1290	59.89	0.049
420	15.99	0.047	860	38.28	0.052	1300	60.38	
430	16.46	0.047	870	38.80	0.052			

APPENDIX I

Table 5

Thermocouple   NiCr–Constantan

Thermal e.m.f. in mV
Temperature of reference sites: 0 °C

°C	mV	mV/°C	°C	mV	mV/°C	°C	mV	mV/°C
0	0	0.063	340	23.79	0.080	680	51.42	0.081
10	0.63	0.062	350	24.59	0.080	690	52.53	0.080
20	1.25	0.061	360	25.39	0.080	700	53.03	0.080
30	1.86	0.061	370	26.19	0.080	710	53.83	0.080
40	2.47	0.059	380	26·99	0.080	720	54.63	0.080
50	3.06	0.060	390	27.79	0.081	730	55.43	0.080
60	3.66	0.062	400	28.60	0.080	740	56.23	0.079
70	4.28	0.063	410	39.40	0.081	750	57.02	0.079
80	4.91	0.064	420	30.21	0.081	760	57.81	0.079
90	5.55	0.066	430	31.02	0.082	770	58.60	0.079
100	6.21	0.065	440	31.84	0.081	780	59.39	0.078
110	6.86	0.066	450	32.65	0.082	790	60.17	0.078
120	7.52	0.066	460	33.47	0.081	800	60.95	0.078
130	8.18	0.068	470	34.28	0.082	810	61.73	0.078
140	8.86	0.068	480	35.10	0.083	820	62.51	0.078
150	9.54	0.070	490	35.93	0.082	830	63.29	0.078
160	10.24	0.070	500	36.75	0.081	840	64.07	0.078
170	10.94	0.071	510	37.56	0.081	850	64.85	0.076
180	11.65	0.071	520	38.37	0.082	860	65.61	0.075
190	12.36	0.072	530	39.19	0.082	870	56.36	0.075
200	13.08	0.074	540	40.01	0.081	880	67.11	0.075
210	12.82	0.074	550	40.82	0.082	890	67.86	0.075
220	14.66	0.074	560	41.64	0.083	900	68.61	0.075
230	15.40	0.074	570	42.47	0.082	910	69.36	0.075
240	16.14	0.075	580	43.29	0.083	920	70.11	0.075
250	16.89	0.075	590	44.12	0.082	930	70.86	0.074
260	17.64	0.075	600	44.94	0.081	940	71.60	0.074
270	18.49	0.075	610	45.85	0.081	950	72.34	0.074
280	19.14	0.075	620	46.56	0.081	960	73.08	0.074
290	19.89	0.076	630	47.37	0.081	970	73.82	0.074
300	20.65	0.078	640	48.18	0.081	980	74.56	0.073
310	21.43	0.078	650	48.99	0.081	990	75.29	0.073
320	22.21	0.079	660	49.80	0.081	1000	76.02	
330	23.00	0.079	670	50.61	0.081			

Table 6

Thermocouple Ag–Constantan

Thermal e.m.f. in mV
Temperature of reference sites: 0 °C

°C	mV	mV/°C	°C	mV	mV/°C	°C	mV	mV/°C
0	0	0.040	210	9.68	0,053	410	21.18	0.062
10	0.40	0.041	220	10.21	0,053	420	21.80	0.063
20	0.81	0.042	230	10.74	0,054	430	22.43	0.063
30	1.23	0.042	240	11.28	0.054	440	23.06	0.064
40	1.65	0.042	250	11.82	0.055	450	23.70	0.064
50	2.07	0.042	260	12·37	0.055	460	24.34	0.065
60	2.49	0.043	270	12.92	0.056	470	24.99	0.065
70	2.92	0.044	280	13.48	0.056	480	25.64	0.066
80	3.36	0.045	290	14.04	0.057	490	26.30	0.066
90	3.81	0.045	300	14.61	0.057	500	26.96	0.066
100	4.26	0.045	310	15.18	0.058	510	27.62	0.067
110	4.71	0.046	320	15.76	0.058	520	28.29	0.067
120	5.17	0.047	330	16.34	0.059	530	28.96	0.067
130	5.64	0.048	340	16.93	0.059	540	29.63	0.068
140	6.12	0.049	350	17.52	0.060	550	30.31	0.068
150	6.61	0.050	360	18.12	0.060	560	30.99	0.069
160	7.11	0.050	370	18.72	0.061	570	31.68	0.069
170	7.61	0.051	380	19.33	0.061	580	32.37	0.069
180	8.12	0.052	390	19.94	0.062	590	33.06	0.070
190	8.64	0.052	400	20.56	0.062	600	33.76	
200	9.16	0.052						

REFERENCES

1. KNIGHT J. R., RHYS D. W. The Platinum Metals in Thermometry Ed. Engelhard Industries Ltd., London, Publ. 2244, (1961)
2. Tables of e.m.f. of Thermoelements. Harens–Hanau.

APPENDIX II

RECOMMENDATIONS OF THE ICTA
STANDARDIZATION COMMITTEE

NOMENCLATURE IN THERMAL ANALYSIS – I
(Reprinted from *Talanta*, **16**, 1227 (1969), by permission)

Nomenclature in thermal analysis is neither uniform nor consistent and can at times be confusing. Because of this a committee consisting of Dr. R. C. Mackenzie (Chairman), Mr. C. J. Keattch (Secretary), Dr. J. P. Redfern and Dr. A. A. Hodgson was appointed at the First International Conference on Thermal Analysis to explore this field. A report, arrived at after consultation with experts in all major English-speaking countries and in other countries interested in this aspect, was approved in principle at the Second International Conference on Thermal Analysis and the Council of the International Confederation for Thermal Analysis (ICTA) have directed that it be published.

This first report is offered as a definitive document of ICTA, the recommendations in which ought to be adhered to in all publications in the English language. It is appreciated that this is only a beginning and that many aspects still require attention; furthermore, new developments in the science may lead to minor revision. Such matters will be the subject of later reports.

Since linguistic considerations render difficult universal application of terms and it may be that names unacceptable in one language are normal usage in another, sub-committees are at present considering the position regarding the French, German, Japanese and Russian languages. The decisions of these sub-committees will be published later and developments in the field of nomenclature will be reported from time to time in the *ICTA Newsletter*.

I. General Recommendations

(*a*) *Thermal analysis* and *not* "thermography" should be the acceptable name in English, since the latter has at least two other meanings in this language, the major one being medical (*Sci. Progr. London*, 1967, 55, 167). The adjective should then be *thermoanalytical* (*cf.* physical chemistry and physicochemical): the term "thermoanalysis" is not supported (on the same logical basis).

(*b*) *Differential* should be the adjectival form of *difference*; *derivative* should be used for the first derivative (mathematical) of any curve.

(*c*) The term "analysis" should be avoided as far as possible since the methods considered do not comprise analysis as generally understood chemically: terms such as *differential thermal analysis* are too widely accepted, however, to be changed.

(*d*) The term *curve* is preferred to "thermogram" for the following reasons:

1. "Thermogram" is used for the results obtained by the medical technique of thermography — see (*a*).
2. If applied to certain curves (*e.g.*, thermogravimetric curves), "thermogram" would not be consistent with the dictionary definition.
3. For clarity there would have to be frequent use of terms such as differential thermogram, thermogravimetric thermogram, *etc*, which are not only cumbersome but also confusing.

(*e*) In multiple techniques, *simultaneous* should be used for the application of two or more techniques to the same sample at the same time: *combined* would then indicate the use of separate samples for each technique.

(*f*) *Thermal decomposition* and similar terms are being further considered by the Committee.

II. Terminology

Acceptable names and abbreviations, together with names which were for various reasons rejected, are listed in Table I. The Committee are in accord with the suggestion, made during discussion of the report, that the limited number of abbreviations considered permissible should be adopted internationally, irrespective of language.

The committee do not wish to pronounce on nomenclature in borderline techniques (such as thermometric titrimetry or calorimetry) which are, to its knowledge, being considered by other bodies. Consideration of techniques not yet extensively employed has been deferred.

Table I.

Recommended terminology

Acceptable name	Acceptable abbreviation*	Rejected name(s)
A. *General*		
Thermal analysis		Thermography
		Thermoanalysis
B. *Methods associated with weight change*		
1. *Static*		
Isobaric weight-change determination		
Isothermal weight-change determination		Isothermal thermogravimetric analysis
2. *Dynamic*		
Thermogravimetry	TG	Thermogravimetric analysis
		Dynamic thermogravimetric analysis
Derivative thermogravimetry	DTG	Differential thermogravimetry
		Differential thermogravimetric analysis
		Derivative thermogravimetric analysis
C. *Methods associated with energy change*		
Heating curves†		Thermal analysis
Heating-rate curve†		Derivative thermal analysis
Inverse heating-rate curve†		
Differential thermal analysis	DTA	Dynamic differential calorimetry
Derivative differential thermal analysis		
Differential scanning calorimetry	DSC	
D. *Methods associated with evolved volatiles*		
Evolved gas detection	EGD	Effluent gas detection
Evolved gas analysis‡	EGA	Effluent gas analysis
		Thermovaporimetric analysis
E. *Methods associated with dimensional change*		
Dilatometry		
Derivative dilatometry		
Differential dilatometry		
F. *Multiple techniques*		
Simultaneous TG and DTA, *etc.*		DATA (Differential and thermogravimetric analysis)
		Derivatography
		Derivatographic analysis

* Abbreviations should be in capital letters without full-stops, and should be kept to the minimum to avoid confusion.

† When determinations are performed during the cooling cycle these become *Cooling curves, Cooling-rate* and *Inverse cooling-rate curves*, respectively.

‡ The method of analysis should be clearly stated and abbreviations such as MTA (mass-spectrometric thermal analysis) and MDTA (mass spectrometry and differential thermal analysis) avoided.

III. Definitions and conventions

A. General

Thermal analysis. A general term covering a group of related techniques whereby the dependence of the parameters of any physical property of a substance on temperature is measured.

B. Methods Associated with Weight Change

1. STATIC

Isobaric weight-change determination. A technique of obtaining a record of the equilibrium weight of a substance as a function of temperature (T) at a constant partial pressure of the volatile product or products.

The record is the isobaric weight-change curve; it is normal to plot weight on the ordinate with weight decreasing downwards and T on the abscissa increasing from left to right.

Isothermal weight-change determination. A technique of obtaining a record of the dependence of the weight of a substance on time (t) at constant temperature.

The record is the isothermal weight-change curve; it is normal to plot weight on the ordinate with weight decreasing downwards and t on the abscissa increasing from left to right.

2. DYNAMIC

Thermogravimetry (TG). A technique whereby the weight of a substance, in an environment heated or cooled at a controlled rate, is recorded as a function of time or temperature.

The record is the thermogravimetric or TG curve; the weight should be plotted on the ordinate with weight decreasing downwards and t or T on the abscissa increasing from left to right.

Derivative thermogravimetry (DTG). A technique yielding the first derivative of the thermogravimetric curve with respect to either time or temperature.

The curve is the derivative thermogravimetric or DTG curve; the derivative should be plotted on the ordinate with weight losses downwards and t or T on the abscissa increasing from left to right.

C. Methods Associated with Energy Change

Heating curves. These are records of the temperature of a substance against time, in an environment heated at a controlled rate.

T should be plotted on the ordinate increasing upwards and t on the abscissa increasing from left to right.

Heating-rate curves. These are records of the first derivative of the heating curve with respect to time (*i.e.*, dT/dt) plotted against time or temperature.

The function dT/dt should be plotted on the ordinate and t or T on the abscissa increasing from left to right.

Inverse heating-rate curves. These are records of the first derivative of the heating curve with respect to temperature (*i.e.*, dt/dT) plotted against either time or temperature.

The function dt/dT should be plotted on the ordinate and t or T on the abscissa increasing from left to right.

Differential thermal analysis (DTA). A technique of recording the difference in temperature between a substance and a reference material against either time or temperature as the two specimens are subjected to identical temperature regimes in an environment heated or cooled at a controlled rate.

The record is the differential thermal or DTA curve; the temperature difference (ΔT) should be plotted on the ordinate with endothermic reactions downwards and t or T on the abscissa increasing from left to right.

Derivative differential thermal analysis. A technique yielding the first derivative of the differential thermal curve with respect to either time or temperature.

The record is the derivative differential thermal or derivative DTA curve; the derivative should be plotted on the ordinate and t or T on the abscissa increasing from left to right.

Differential scanning calorimetry (DSC). A technique of recording the energy necessary to establish zero temperature difference between a substance and a reference material against either time or temperature as the two specimens are subjected to identical temperature regimes in an environment heated or cooled at a controlled rate.

The record is the DSC curve; it represents the amount of heat applied per unit time as ordinate against either t or T as abscissa.

D. Methods Associated with Evolved Volatiles

Evolved gas detection (EGD). This term covers any technique of detecting whether or not a volatile product is formed during thermal analysis.

Evolved gas analysis (EGA). A technique of determing the nature and/or amount of volatile product or products formed during thermal analysis.

E. Methods Associated with Dimensional Change

Dilatometry. A technique whereby changes in dimension(s) of a substance are measured as a function of temperature.

The record is the dilatometric curve.

Derivative dilatometry; differential dilatometry. These terms carry the connotations given in I(*b*) above.

F. Multiple Techniques

This term covers simultaneous DTA and TG, etc., and definitions follow from the above.

IV. Acknowledgements

The Committee express their thanks to the Society for Analytical Chemistry and the Thermal Analysis Group of that Society for assistance rendered and also to thermal analysis in many countries for detailed comments at various stages of the programme.

The Macaulay Institute for Soil Research   R. C. MACKENZIE
Craigiebuckler   *Chairman, ICTA Nomenclature Committee*
Aberdeen
Scotland

NOMENCLATURE IN THERMAL ANALYSIS – II
(Reprinted from *Talanta*, 1972, **19**, 1079, by permission)

The recommendations in the First Report of the Nomenclature Commitee of the International Confederation for Thermal Analysis (ICTA) have been generally well received and are at present being considered for adoption by both the International Union of Pure and Applied Chemistry and the International Standards Organization. Considerable interest has also been shown by individual scientists.[2]

Since these proposals were promulgated, the Committee have drawn up a further report, the recommendations in which have been endorsed in Business Session at the Third International Conference on Thermal Analysis at Davos, Switzerland, in August, 1971. The Council of ICTA have therefore directed that this Second Report be also published as a definitive document of ICTA, with the recommendation that the conventions set out therein be adhered to in all publications in the English language. Background information on the reasons for adoption of certain conventions has already been published.[3]

The Sub-Committees dealing with the German, Japanese and Russian languages[1] have all been active and a nomenclature system in Japanese, based on the recommendations in the First Report, has now been published.[4]

I. Amplification of first report

Because of the variety of opinions expressed on, and the different interpretations of, the term *pyrolysis*, the Committee consider the time inopportune to promulgate on *thermal decomposition* and related terms—see First Report, Section I (*f*).

Some confusion appears to have arisen over the term *isobaric weight-change determination* as defined in the First Report, Section IIIB, Sub-section 1. The Committee consider that this confusion could be obviated by the following statement.

> In the context of the report published in *Talanta*, **16**, 1227, (1969), Section IIIB, Sub-sections 1 and 2 headed, respectively, *Static* and *Dynamic*, these terms refer to environmental temperature. It should be noted that the same terms are also used with reference to environmental atmosphere.

II. DTA and TG apparatus and technique

In considering the terms available, certain arbitrary choices have had to be made—*e.g.*, between *specimen* and *sample*—but the only term in fairly common use that is rejected is *inert material*.

A. DTA

The *sample* is the actual material investigated, whether diluted or undiluted.

The *reference material* is a known substance, usually inactive thermally over the temperature range of interest.

The *specimens* are the sample and reference material.

The *sample holder* is the container or support for the sample.

The *reference holder* is the container or support for the reference material.

The *specimen-holder assembly* is the complete assembly in which the specimens are housed. Where the heating or cooling source is incorporated in one unit with the containers or supports for the sample and reference material, this would be regarded as part of the specimen-holder assembly.

A *block* is a type of specimen-holder assembly in which a relatively large mass of material is in intimate contact with the specimens or specimen holders.

The *differential thermocouple** or *ΔT thermocouple** is the thermocouple* system used to measure temperature difference.

B. TG

A *thermobalance* is an apparatus for weighing a sample continuously while it is being heated or cooled.

The *sample*† is the actual material investigated, whether diluted or undiluted.

The *sample holder* is the container or support for the sample.

C. DTA and TG

The *temperature thermocouple** or *T thermocouple** is the thermocouple* system used to measure temperature; its position with respect to the sample should always be stated.

The *heating rate* is the rate of temperature increase, which is customarily quoted in degrees per minute (on the Celsius or Kelvin scales). Correspondingly, the cooling rate is the rate of temperature decrease. The heating or cooling rate is said to be *constant* when the temperature time curve is linear.

In simultaneous DTA-TG, definitions follow from those given for DTA and TG separately.

III. DTA and TG curves

The Committee, in reaffirming their decision to use the terms *differential thermal curve* or *DTA curve, thermogravimetric curve or TG curve*

---

* Should another thermosensing device be used, its name should replace *thermocouple* in these terms.

† Samples used in TG are normally not diluted, but in simultaneous TG and DTA diluted samples might well be used.

and *derivative thermogravimetric curve* or *DTG curve*, recommend disuse of the other terms which have appeared in the literature, such as thermogravimetric analysis curve, thermolysis curve, thermoweighing curve, thermogravigram, thermoponderogram, thermogram, differential thermogravimetric curve, differential thermogram, derivative thermogram, polytherm, *etc.*

Certain conventions and reporting procedures for DTA and TG curves have already been specified in *Anal. Chem.*, **39**, 543, (1967), and in the First Report. The following definitions are to be read in conjunction with these recommendations.

A. DTA

In DTA *it must be remembered* that although the ordinate is conventionally labelled ΔT the output from the ΔT thermocouple will in most instances vary with temperature and the measurement recorded is normally the e.m.f. output, E, *i.e.*, the conversion factor, b, in the equation $\Delta T = bE$ is not constant since $b = f(T)$, and that a similar situation occurs with other sensor systems.

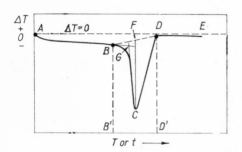

Fig. A.1 Formalized DTA curve.

All definitions refer to a single peak such as that shown in Fig. A1: multiple peak systems, showing shoulders or more than one maximum or minimum, can be considered to result from superposition of single peaks.

The *base line* (*AB* and *DE*, Fig. A.1) corresponds to the portion or portions of the DTA curve for which ΔT is approximately zero.

A *peak* (*BCD*, Fig. A.1) is that portion of the DTA curve which departs from and subsequently returns to the base line.

An *endothermic peak* or *endotherm* is a peak where the temperature of the sample falls below that of the reference material; that is, ΔT is negative.

An *exothermic peak* or *exotherm* is a peak where the temperature of the sample rises above that of the reference material; that is, ΔT is positive.

Peak width (*B'D'*, Fig. A.1) is the time or temperature interval between the points of departure from and return to the base line.*

---

* There are several ways of interpolating the base line and that given in Fig. A.1 is only an example. Location of points *B* and *D* (Fig. A.1) depends on the method of interpolation of the base line.

Peak height (*CF*, Fig. A.1) is the distance, vertical to the time or temperature axis, between the interpolated base line* and the peak tip (*C*, Fig. 1).

Peak area (*BCDB*, Fig. A.1) is the area enclosed between the peak and the interpolated base line.*

The *extrapolated onset* (*G*, Fig. A.1) is the point of intersection of the tangent drawn at the point of greatest slope on the leading edge of the peak (*BC*, Fig. 1) with the extrapolated base line (*BG*, Fig. A.1).

B. TG

All definitions refer to a single-stage process such as that shown in Fig. A.2: multistage processes can be considered as resulting from a series of single-stage processes.

A *plateau* (*AB*, Fig. A.2) is that part of the TG curve where the weight is essentially constant.

The *initial temperature*, T_i, (*B*, Fig. A.2) is that temperature (on the Celsius or Kelvin scales) at which the cumulative weight change reaches a magnitude that the thermobalance can detect.

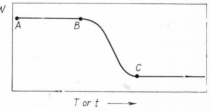

Fig. A.2 Formalized TG curve.

The *final temperature*, T_f, (*C*, Fig. A.1) is that temperature (on the Celsius or Kelvin scales) at which the cumulative weight change reaches a maximum.

The *reaction interval* is the temperature difference between T_i and T_f as defined above.

IV. Acknowledgements

The Committee express their thanks to the Society for Analytical Chemistry for providing accommodation for meetings and secretarial facilities, to the Thermal Analysis Group of that Society for assistance rendered, and to thermal analysts in many countries for cooperation in providing comments at various stages of the programme.

c/o Industrial and Laboratory Services
Lyme Regis
Dorset, U. K.

R. C. MACKENZIE
C. J. KEATTCH
D. DOLLIMORE
J. A. FORRESTER
A. A. HODGSON
J. P. REDFERN

REFERENCES

1. *Talanta*, **16**, 1227 (1969).
2. *Chem. Ind.*, 1970, 272, 449, 515, 643; 1917, 57.
3. R. C. MACKENZIE, *J. Therm. Anal.*, **4**, 215 (1972).
4. *Calorim. Therm. Anal. News Letter.*, **2**, 45 (1971).

APPENDIX III

REFERENCES TO DTA MICROMETHODS

1. AKIRA YAMAMOTO, KIYOTSUGU YAMADA, MICHIO MARUTA, JUNICHI AKIYAMA. *A New Design of Micro Sample DTA Apparatus and its Application*, in *Thermal Analysis*, Vol. 1, Ed. R.F. Schwenker Jr., P. D. Garn, Academic Press. New York 1969, p. 105.
2. ADAM W., SANABIA J. *DSC Applied to the Thermolysis of Peroxalates*, in *Abstracts of Papers of the 3rd Intern. Conf. on Thermal Analysis*, Davos 1971, No. II—21.
3. ADLER P., GESCHWIND G., DELASI R. *Calorimetric Study of Precipitation in a Commercial (7075) Aluminium Alloy*, in *Abstracts of Papers of the 3rd Intern. Conf. on Thermal Analysis*, Davos 1971, Nr. II-64.
4. ARSENEAN D. F., STANWICK J. J. J. *A Study of Reaction Mechanisms by DSC and TG*, in *Abstracts of Papers of the 3rd Intern. Conf. on Thermal Analysis*, Davos 1971.
5. BARRALL E. D. II., PORTER R. S., JOHNSON J. F. *Microboiling Point Determination at 30 to 760 torr by DTA*. *Anal. Chem.* **37**, 1053 (1965).
6. BAXTER R. A. *A Scanning Microcalorimetry Cell Based on a Thermoelectric Disc—Theory and Application*, in *Thermal Analysis*, Vol. 1, Ed. R. F. Schwenker Jr. P. D. Garn, Academic Press, New York 1969, p. 65.
7. BLOCK J. *Use of DSC for the Analysis of Chloride-Bromide Mixtures*, *Anal. Chem.* **37**, 1414 (1965).
8. BOSIO L. *Surfusion et polymorphisme du gallium à la pression atmospherique. Metaux* **40**, 481 (1965).
9. BOSIO L., DEFRAIN C. R. *Compt. Rend.* **258**, 4929 (1963).
10. BURR J. T. *The Application of DTA at Constant Temperature to Evaluate Hazardous Thermal Properties of Chemicals*, in *Thermal Analysis*, Vol. 1, Ed. R. F. Schwenker Jr, P. D. Garn, Academic Press, New York 1969, p. 301.
11. BOHON R. L. *Approximate Heats of Explosion Using Differential Thermal Analysis*, *Anal. Chem.* **35**, 1845 (1963).
12. CARROLL R. W., MANGRAVITE R. V. *Simultaneous Scanning Calorimetry and Conductivity*, in *Thermal Analysis* Vol. 1., Ed. R. F. Schwenker Jr., P. D. Garn, Academic Press, New York 1969, p. 189.
13. CURRELL B. R. *Application of DTA to the Quantitative Measurements of Enthalpy Changes*, in *Thermal Analysis*, Vol. 2., Ed. R. F. Schwenker Fr., P. D. Garn, Academic Press, New York 1969, p. 1185.
14. CRIGHTON J. S., HOLMES D. A. *The Characterisation of Textile Fibre Blends Containing Polyamide by DTA*, in *Abstracts of Papers of the 3rd Intern. Conf. on Thermal Analysis*, Davos 1971, No. III-37m.
15. CHIU J. *Visual Observation in DTA*, *Anal. Chem.* **35**, 933 (1963).
16. DAVID D. J. *Determination of Specific Heat and Heat of Fusion by DTA. Anal. Chem.* **36**, 2162 (1964).

17. DUSOLLIER G., ROBREDO J. *Applications de l'analyse thermique differentielle à l'étude de certaines variations de composition dans le verre, Verres et Refr.* **21,** 550 (1967).

18. DUSOLLIER G., ROBREDO J. *Étude par ATD et par microscopie de la dévitrification de quelques verres à cristallisation rapide, Verres et Refr.* **23,** 10 (1969).

19. DURUZ, J. J. TISSOT P., MONNIER R. *Un nouveau dispositif de micro-analyse thermique. Helv. Chim. Acta* **50,** 822 (1967)

20. *Du Pont Thermogram* **2** (3) (June 1965).

21. FREEBERG F. E., ALLEMAN F. G. *A Sealed Cell for Use with a Commercial Differential Scanning Calorimeter. Anal. Chem.* **38,** 1806 (1966).

22. HIROTARO KAMBE, ITARU MITA, KAZUYUKI HORIE. *Kinetic Investigation of Polymerisation Reaction with DSC,* in *Thermal Analysis,* Vol. 2, Ed. R. F. Schwenker Jr., P. D. Garn, Academic Press, New York 1969 p. 1071.

23. HIROTARO KAMBE, ITARU MITA, RIKIO YOKOTA, *Melting of some Aromatic and Heterocyclic Oligomers,* in *Abstracts of Papers of the 3rd Intern. Conf. on Thermal Analysis,* Davos 1971, No. III-35.

24. JAFFE M., WUNDERLICH B. *Superheating of Extended-Chain Polymer Crystals,* in *Thermal Analysis* Vol. 1, R. F. Schwenker Jr., P. D. Garn, Academic Press, New York 1969, p. 387.

25. JUNICHI AKIYAMA, KALSUYA SATO. *Theoretical Consideration of Micro Sample DTA,* in *Abstracts of Papers of the 3rd Intern. Conf. on Thermal Analysis,* Davos 1971, No. I-4.

26. KRAWETZ A. A., TOVROG T. *Determination of Vapor Pressure by Differential Thermal Analysis, Rev. Sci. Instr.* **33,** 1465 (1962).

26a KING W. H. JR., CAMILLI C. T., FINDEIS A. F. *Thin Film Thermocouples for DTA, Anal. Chem.* **40,** 1330 (1968).

27. LEVY P. F., FITZPATRICK W. J. *Pittsburgh Conf. Anal. Chem. Appl. Spectroscopy,* Pittsburg, March 5–10, 1967, Paper 68.

28. LOCKE C. E., STONE R. L. *Study of Dehydration of Magnesium Sulfate Heptahydrate,* in *Thermal Analysis* Vol. 2, Ed. R. F. Schwenker Jr., P. D. Garn, Academic Press New York 1969, p. 963.

29. LINSEIS M. *Eine neue DSC Methode,* in *Abstracts of Papers of the 3rd Intern. Conf. on Thermal Analysis,* Davos 1971, No. I–23.

30. MAZIÈRES Ch. *Microanalyse thermique differentielle, Applications Physicochimiques. Bull. Soc. Chim. France.* 1695 (1961).

31. MAZIÈRES Ch. *Micro and Semimicro Differential Thermal Analysis, Anal. Chem.* **36,** 602 (1964).

32. MOORE R. *DTA of Suspension of Human Erythrocytes and Glycerol,* in *Thermal Analysis* Vol. 1, Ed. R. F. Schwenker Jr., P. D. Garn, Academic Press New York 1969, p. 615.

33. NEUMANN G., COLLINS W. E. *The Evaluation of Catalysts by Pressure DSC,* in *Abstracts of Papers of the 3rd Intern. Conf. on Thermal Analysis,* Davos 1971, No. I-12.

34. O'NEILL M. J. *The Analysis of a Temperature-Controlled Scanning Calorimeter, Anal. Chem.* **36,** 1238 (1964).

35. O'NEILL M. J. *Measurements of Specific Heat Functions by Differential Scanning Calorimetry, Anal. Chem.* **38,** 1331 (1966).

36. O'NEILL M. J., WATSON E. S., JUSTIN J., BRENNER N., GRAY A. P. *Abstr. Pittsburgh Conf. Anal. Chem. Appl. Spectroscopy* No. 62, p. 60, March 4–8 (1963).

37. PRENDERGAST J. A. *The Rapid Determination of Solid Index with a Scanning Calorimeter*, in *Thermal Analysis* Vol. 2, Ed. R. F. Schwenker Jr., P. D. Garn, Academic Press, New York 1969, p. 1317.

38. REID D. S. *The Sol-Gel Transition in Polysaccharide Gels*, in *Abstracts of Papers of the 3rd Intern. Conf. on Thermal Analysis*, Davos 1971, No. III-38.

39. SOMMER G., SANDER H. W. *Balanced Temperature Analysers*, in *Thermal Analysis* Vol. 1, Ed. R. F. Schwenker Jr., P. D. Garn, Academic Press, New York 1969, p. 163.

40. STONE R. L. *DTA by the Dynamic Gas Technique*, Anal. Chem. **32**, 1582 (1960).

41. SPEROS D. M., WOODHOUSE R. L. *Quantitative DTA. Heat and Rates of Solid-Liquid Transitions*, J. Phys. Chem. **67**, 2164 (1963).

42. VAN TETS A., WIEDEMANN H. G. *Simultanous Thermomicroscopic and Differential Thermal Investigations of Melting and Freezing Processes*, in *Thermal Analysis* Vol. 1, Ed. R. F. Schwenker Jr., P. D. Garn, Academic Press, New York 1969, p. 121.

43. WITTELS M. *The Differential Thermal Analyser as a Micro Calorimeter*, Am. Mineralogist **26**, 615 (1951).

44. WATSON E. S., O'NEILL M. J., JUSTIN J., BRENNER N. *A Differential Scanning Calorimeter for Quantitative Differential Thermal Analysis*, Anal. Chem. **36**, 1233 (1964).

45. WISNEWSKI A. M., CALHOUN R. J. JR., WITNAUER L. P. *Pressure Chamber for Use During DTA*, J. Appl. Polymer Sci. **9**, 3935 (1965).

46. VASSALLO D. A., HARDEN J. C. *Precise Phase Transition Measurements of Organic Materials by DTA*, Anal. Chem. **34**, 132 (1962).

47. WUNDERLICH B., BOPP R. C. *DSC at 150 Atm. Pressure*, in *Abstracts of Papers of the 3rd Intern. Conf. on Thermal Analysis* Davos 1971, No. I-25.

48. YAMAMOTO A., AKIYAMA J., OKINO T. *Micro Differential Thermal Analysis of Water of Crystallization in Strontium Chloride*, Chem. Anal. **17**, 1126 (1968).

49. YAMAMOTO A., YAMADA K., AKIYAMA J., OKINO T. *Micro Differential Thermal Analyzer*, J. Chem. Soc. Japan, Ind. Chem. Sec. **71**, 2020 (1968).

50. ROSICKÝ J. *Přístroj pro DTA s vyměnitelnými hlavicemi*, Silikáty **12**, 295 (1968).